Ceramics and Print

Free at Last by Paul Scott (UK), height 30 cm, 1990, Porcelain form, transparent glaze, inglaze screen printed decals, 1200°C, lustres. *Photograph by Andrew Morris.*

CERAMICS AND PRINT

Second edition

Paul Scott

A & C Black • London
University of Pennsylvania Press • Philadelphia

First published 1994

This second edition first published in
Great Britain 2002
A&C Black (Publishers)
37 Soho Square
London W1D 3QZ

ISBN 0 7136 5485 6

Published simultaneously in the USA by
University of Pennsylvania Press
4200 Pine Street
Philadelphia, PA 19104-4011

ISBN 0-8122-1800-0

A CIP catalogue record for this book is available from the British Library and the US Library of Congress.

Cover illustrations
front: Detail from a *Cumbrian Blue(s)* plate, Paul Scott (UK) 1998. Inglaze decal screen print collage on bone china plate. *Photograph by Andrew Morris.*

back: Detail from Dick Lehman's *Flattened Bottle with Handles*, 25.5 cm (10") tall, thrown and altered grolleg porcelain, saggar-fired kiln print with sumac and roadside grasses. *Photograph by Dick Lehman.*

Contents

Acknowledgements

Thanks to everyone who has given freely of their time, provided information and photographs. In addition, special thanks to the following for a variety of reasons: inspiration, help, or for being there when I needed them...

For the first edition of *Ceramics and Print*: Adrian Bailey, (the late) Kenneth Beaulah, Jim Bennett, John Calver, Tim Challis, Robert Copeland, Graeme Cruikshank, Malcolm Gibson, Barry Gregson, (the late) Vronwy Hankey, Paul and Audrey Kettle, Linda McRae, Andrew Morris, Beverly Nenk, Paul Norton, Matthew Rice, Richard Slee, Jane Smith, Amanda Spencer-Cooke, Alan Strogen, Mike Verden.

For this updated version, Steve Hoskins, Pat King, Les Lawrence, Scott Rench and Rimas Visgirda.

Thanks to Anne and Ellen, and Linda Lambert my Editor.

Finally, this update has been made possible not only because of existing contacts and networks, but also because individuals have contacted me about their discoveries and new work. If you wish to be considered for inclusion in future updates, please email me at ceramicsandprint@mac.com

Paul Scott May 2001.

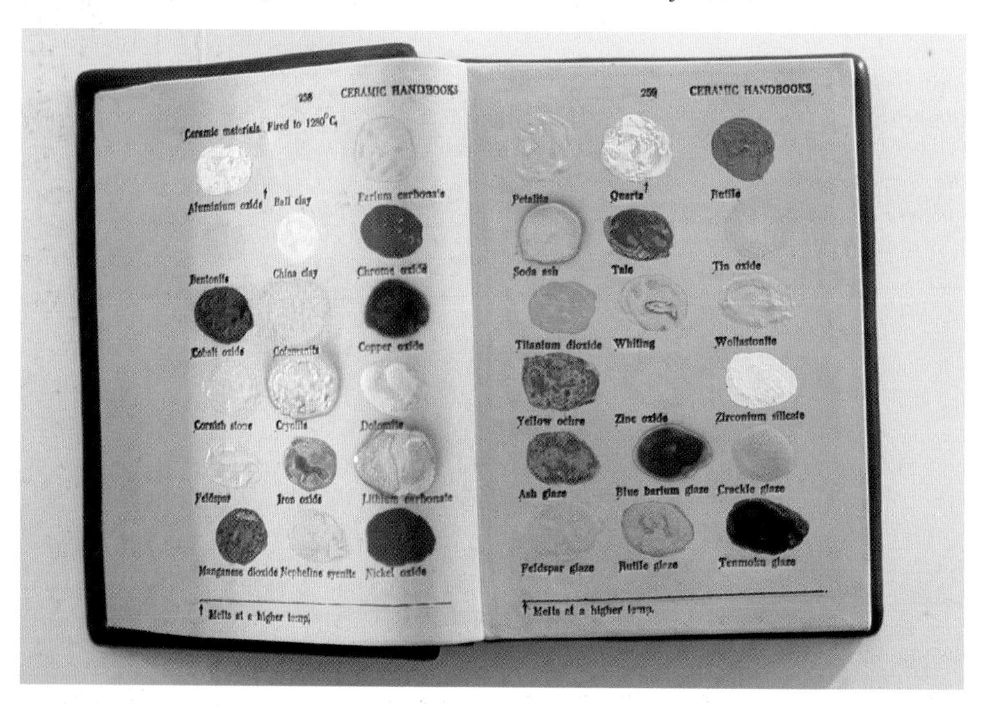

Glaze Book, Lenny Goldenberg (Denmark), approximately 30 cm in length, a press-moulded, hand-built and impressed ceramic book from the 'Keramisk Handbog' series. Goldenberg has used a variety of print equipment normally associated with the newspaper printing industry in his work.

Introduction

Ceramics and Print have in common the ability to repeat a shape, form or image and have been used together for hundreds of years to produce decorative wares and tiles. Although ceramics are perceived as primarily three-dimensional, and print two-dimensional, ceramic surfaces are variously both.

A decade ago, to find print in a piece of studio ceramics was a relatively rare occurrence... but such a dramatic transformation in attitudes has taken place in recent years, that now, printing has become recognised as 'an integral part of current studio practice'.[1&2] Similarly but perhaps less dramatically, in printmaking studios, the advent of paperclay and a continuing interest in multimedia prints has led to a gradual exploration of the possibilities that ceramic materials offer.

The first edition of this little handbook has been credited by some as being a catalyst for the explosion in print practice in ceramics studios, but in truth the connections were already there, the book just helped the explorations along the way.

Much of what appeared in the original edition of *Ceramics and Print* has not dated, but what has happened is a diversification and proliferation of process and intent. From decorative icons on slip cast ceramics sold in craft shops, experimental one off prints in print studios, limited edition inglaze screen print decals on plates, large-scale commissioned work for public places, to conceptual pieces relying on the printed image and the ceramic object's connotations to work, the material connectivity between Ceramics and Print is producing experimentation and innovation.

Teas No. 7, by Paul Scott (UK), 1977, Screen printed tiles 15cm x 45cm (6"x18"). Printed with reactive glazes onto biscuited dry pressed tile. *Photograph by Barry Gregson.*

[1&2] *Ceramics in Society Magazine* No 37, Autumn 1999, p. 6.

My first encounter with print and ceramics came in my final year at college. As an escapee from the painting studio, I was more interested in ceramic surfaces than form, so my work in the ceramics department tended towards the two-dimensional. After a visit to Johnstone's tile factory in Stoke on Trent where I saw industrially produced screen-printed tiles, I came away with samples of reactive glazes and blank bisque tiles.

In the studio, I painted an image in varnish onto a silk screen, and printed through the screen with ceramic colours in an oil-based medium onto tiles. The varnish acted to block the screen so ink appeared in the areas that had not been painted. The medium, however, began to dissolve the varnish, and so the image degraded. Unintentionally, I ended up with a large panel where a series of images of a derelict building, progressively disintegrated. I remember quite clearly the sense of excitement generated as, after the squeegee was pulled across the screen, I lifted it to reveal the printed image on the tiles underneath. This sense of excitement, as images are revealed following the print process, is part of the fascination with printmaking in whatever form it takes, but in the case of a ceramic 'print' it is further heightened, by the additional process of fire that it is subject to. Every ceramist knows that whatever the type of kiln, however predictable the glaze and colours, however many times one fires, the glaze firing presents a period of excitement and anticipation. You are never exactly sure of what will present itself when you open the kiln door. In the case of the original tiles I printed, there was the 'magic' of the reactive glazes producing the variety of colours at the points where the screened image met the overlying glaze.

Detail of *A Residency at Whitehaven School*, Paul Scott (UK), 1986.Ceramic relief panel, painted underglazes with photocopied text, postcards etc., 59 cm x 42 cm (23"x16.5").

Years later when I was collaging paper images onto a fired ceramic surface, I became dissatisfied with the mixed media I was being forced to adopt. I remembered the college tiles, saw commercially produced pottery clearly showing printed imagery, and set about finding out how to put printed images onto a ceramic surface that was not necessarily flat.

I found that there were limited print products available from ceramic suppliers (certainly no ceramic ones from fine art print suppliers) and no instructions of how to use the materials came with them. A further look at specialist ceramics, then print handbooks, suggested that the two could not meet. Part of the problem it seems has been the historical avoidance of the ceramic by fine art printmakers, and then the insistence by writers on ceramics focusing on form. For studio pottery the ceramic surface has always been a secondary consideration. Typical was Bernard Leach's view in *A Potter's Book:* 'well painted pots have a beauty of expression greater than pottery decorated with engraved transfers, stencils or rubber stamps'. His only other references to printing are equally dismissive: Rubber Stamps 'vulgar patterns, and mechanised shapes and finish, the effect is deplorable'. Of course for the Leach aesthetic, this is an entirely understandable position. The problem has been the wholehearted embrace of all things Leach-like by the contemporary studio pottery movement for a long while to the exclusion of much else.

The use of print-related techniques in themselves do not necessarily lead to 'good' or 'bad' work, it is the way a technique is used, and the context in which the work is made or exhibited which are the key factors. Because a pot is sponge decorated it isn't automatically mechanical and lifeless, much as a painted pot is not automatically beautiful and expressive. As importantly too, print on or in the ceramic surface need not be on a three-dimensional form at all.

The association with 'industry' has also not favoured the ceramic printer in the eyes of the contemporary studio pottery establishment. Ironic, for although the Industrial Revolution did allow for pottery production on a huge scale, and division of labour and skills, pottery production itself was probably one of the first manufacturing processes to be 'industrialised'. Processes of preparation, throwing, decorating and firing were easily identifiable and separated, so specialist tasks could be the exclusive domain of trained craftspeople. In Crete, ceramics from Minoan and Mycenaean civilisations clearly show that pottery was produced on a vast scale, in some sort of 'industrial' way[3]. Vast numbers of crude drinking cups found at numerous Mycenaean archaeological sites suggest the first 'disposable' cup. Repetition throwing itself was most probably the start of the industrialisation of pottery making, but it is a process to this day that is regarded with awe and fascination and a degree of reverence not afforded to other ceramic processes or disciplines.

The potter can use throwing, casting, jigger and jollying, transfers, and the printmaker can use print in purely commercial terms – mass producing forms or images as a business, where pots or prints are just a commodity. But clearly it doesn't have to be this way. Increasingly ceramists use other and more recent 'industrial' processes in creative ways.

Sasha Wardell's slip cast forms are a good example: Using industrial methods to produce work where a degree of repetition is possible (the form), but where the development of surface line and pattern is individual to each piece. Contemporary ceramics has become much more inclusive and open minded, acknowledging that the view Leach provided, although intellectually sound in its way, was a very selective one.

For many years, the artist printmaker has used industrial printing processes for the creative possibilities that print processes provide. Gabor Peterdi in the introduction to his comprehensive manual on printmaking sums it up perfectly:

[3] Personal communication with Vronwy Hankey 1994.

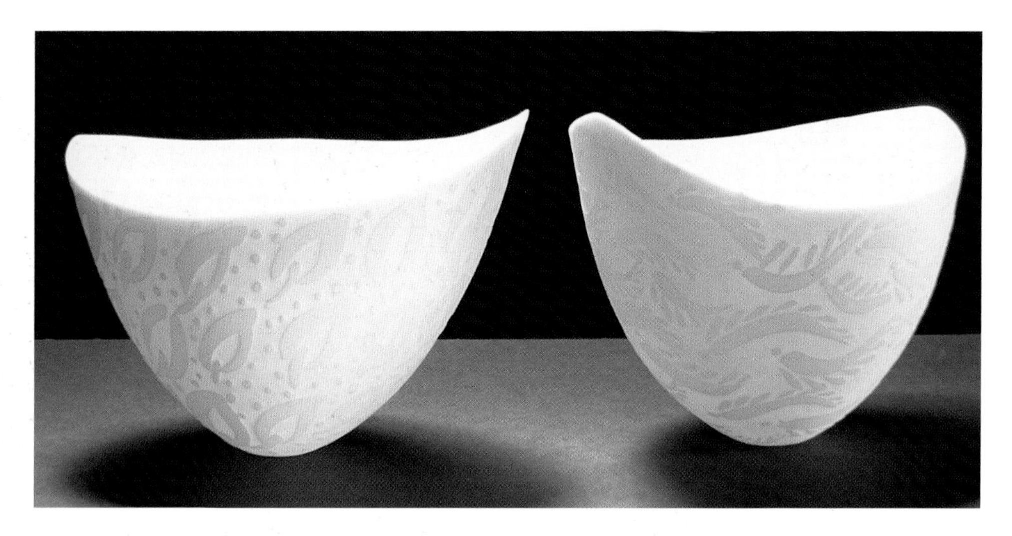

Bowls, by Sasha Wardell (UK). Slip cast, bone china bowls using water erosion technique. Liquitex (a US trade name for medium for acrylic paints, an *acrylic polymer emulsion*) is applied to the bone dry piece after de-moulding and then a damp sponge is used all over the piece several times, eroding the unprotected areas of dry clay until a relief appears. The method of production using slip casting, high bisque and low glaze fire, are both common methods used in industry, but here used to make individual pieces of work.

Restless Sex c.1967, overglaze decal collage on porcelain plate 10.5" dia. *Photo Eduardo Calderon, courtesy Garth Clark Gallery New York USA.* In the 1960s Howard Kottler (USA) collaged open stock industrial decals on ready made bought plates. At the time they raised important questions about appropriation, art and technology, originality and academic hierarchies. (See Failing, Patricia, *Howard Kottler Face to Face*, University of Washington Press 1995.)

Detail platter 56cm (22") dia., John Calver (UK). Stamped, trailed and poured decoration, with slips and glazes. Stoneware, oil kiln, reduction fired.

> *I make prints because in using the metal, the wood, and other materials available, I can express things that I cannot express by any other means. In other words I am interested in printmaking not as a means of reproduction but as an original creative medium. Even if I could pull only one print from each of my plates I would still make them.*

This approach has now seeped into the ceramics studio, and the print materials available to the fine art printmaker have begun to include paperclay, porcelain and ceramic inks.

There are rich and varied traditions of printmaking and ceramics, and use of the processes and connections by contemporary artists is more widespread than ever:[4] Potters like John Calver(UK)

Detail *Fiction*, laser print on tiles, Robert Dawson (UK) 2001.

produce functional domestic ware with complex printed surfaces; Robert Dawson(UK) subverts the traditions of ceramic surface decoration and our perceptions of it with his plates and tiles, whilst Scott Rench (USA) screen prints computer generated images of his personal reflections on life onto large clay slabs, and Richard Shaw's sculptural, surreal *trompe l'oeil* ceramics (see p. 100) rely on the printed surface for much of their authenticity. For contemporary artist Conrad Atkinson the print's value on his plates, ceramic landmines, water closets and sinks is not primarily in the skill of application, or execution, or in the beauty or otherwise of the object that

[4] See Ed Scott Bennett, *Hot off the Press* (Bellew Publishing), 1996; *Glazed Expressions* catalogue, Orleans House, Twickenham, 1998.

I sing my song in hopes that you will sing it back (study) Scott Rench (USA) 2001. Printed underglaze directly onto slip cast earthenware 38cm x 38cm x 23cm (15"x 15"x9") 1100°C (Cone 03). 'I have been playing with different ways of expressing this idea and it has taken many different forms along the way'.

Conrad Atkinson (UK/USA), *Mining Culture, Valmara Willow Pattern Land Mine*. Overglaze open stock decal on earthenware slip cast Land Mine, made for *Transient* exhibition, 1996.

supports it, but in the context in which it is used and viewed. For Atkinson, 'the medium has always been inextricably linked to the purpose of the work, and never its reason for being'.[5]

[5] See Lambert, Susan from *The Victoria and Albert Museum, Conrad Atkinson's Newspaper works from 6 April 1989*, in *Transient*, 'Tullie House Museum and Art Gallery', 1996.

Whilst attitudes to processes may be different and individuals draw on different traditions, they still use common materials and ceramic processes.

In defining the boundaries of the ceramics and print this book examines, I have limited them to those areas not previously documented in detail. So, although moulds, stamps and impressed decoration are considered in a number of contexts, they are not examined in depth. I have rather, taken as a general definition the word 'print' to mean the transfer of ceramic colour from a plate (metal, plaster, wood, lino) or a screen directly or indirectly. In addition direct photographic processes, computer-generated, laser printed and photocopier based prints are considered too.

Some have complained that the original edition of *Ceramics and Print* gave a tantalising glimpse of what was possible but didn't give the exact route to follow to get the results. But it was never my intention to give a step by step account of all the processes involved, instead my purpose was to describe the basic and illustrate the possible, the hope being that the information would engender exploration and innovation. The same model has been used again.

Some might wish that this follow up would contain more technical details, for example, about particular inks for particular processes, but on the whole it doesn't. Previously documented processes are elaborated on by reference to artists' work not available for the first edition, and there are many new photographs and details of working processes. New adaptions and innovations are included, including an extended section on the role of computers, laser printers and photocopiers. But as before, references in the Bibliography guide the reader to more detailed books on printmaking and general ceramic processes. Health and safety issues are addressed in a separate section and listings include suppliers of services, tools and materials and some websites.

There is a need for more detailed documentation on ceramics and screen printing, etching, relief printing, lithography and the possibilities that digital technology offers, but this little handbook is not big enough. Other, more specific and detailed publications are now needed to document the knowledge that this field of study is generating.

In spite of this, and although ceramics and print lack a traditional grounding in the academies of fine arts or any long-term established studio practice, there are actually few rigid rules or procedures; a little knowledge, an enquiring mind and a willingness to explore can produce exciting results. With this small handbook, you will have far more to go on than I did when I first started my investigations...

Chapter One
Historical and industrial background

A knowledge of past processes can help in the understanding of current practice, and the development of new methods and techniques. In particular this area has been little documented in terms accessible to the artist, or the collector of studio ceramics or print. The purpose of this chapter is to look at the historical background of printing and ceramics, and necessarily, its use in industry.

The modern ceramics industry is more and more dependent on high tech, expensive machinery for all its manufacturing processes. It hasn't always been this way, and a look at some of the printing methods (some long since discarded) that have been used by industry, reveals processes readily adaptable for small-scale production, and for creative exploration.

Plastic clay has many properties, not least among them the ability to faithfully record impressions in its surface – from the earliest potter's thumb print, to simple patterned designs stamped into its surface. Potters were perhaps the world's first 'printers'.

The first use of 'print' in the context of transferring a ceramic colour from one body to another (other than painting) appears to be the use of natural sponges to decorate Minoan and Mycenaean pottery. Referred to as 'splatterware' by some archaeologists, its use was said to be to imitate the decoration on ostrich eggs. Whatever the reason, the sponges were used in a relatively random way to produce a mottled effect in contrasting slip on a clay body. The decoration was most commonly used on bottle and cup forms. (The process reappeared with mottled greens and browns under a transparent lead glaze in the middle of the last century.)

Small Jar of the Mycenaean (late Helladic IIIA1) period, c. 1390 to 1370/60, height 10.2 cm (4"). Fine buff clay, shiny slip, dark brown decoration. From a tomb near Khalkis, Euboea, Greece, Khalkis Museum Inventory no. 471. *Photograph by Vronwy Hankey, published with kind permission of the Managing Committee of the British School at Athens, and Dr E. Sakellarakis, Ephor of Euboea.*

Fired clay beaker, 23 cm tall, made approximately 1800BC found at Hunsonby in Cumbria, England. Decorated with patterned, impressed lines, possibly by notched wood or bone implement. *Collection of Tullie House, Carlisle Museum and Art Gallery.*

Relief printing and intaglio, the invention of the transfer

For centuries the only other printing on clay appears to have been the use of wooden, clay or other stamps, natural or man made, to decorate by impressing into the clay surface. Stamps facilitated the simple repetition of shapes creating patterns and designs.

The link between impressing, and printing using the stamp or sponge to transfer colour does not appear to have been made for centuries. Medieval tiles appear to show this as the next development in print. In between the 13th and 16th centuries decorated floor tiles provided rich patterned floors in royal palaces, and ecclesiastical and monastic buildings. They were essentially red clay tiles with painted or inlayed white/buff clay patterns or designs, showing foliage, geometric and heraldic images. It is obvious from the style and quantity of tiles that many were made using wooden stamps or moulds. A contrasting clay in plastic or slip form was applied into the indented areas of the stamped or moulded tile. When the tile was dry, the surface was carefully scraped back and the stamped design appeared, usually white against a red clay background.

It is known that woodcuts, the earliest type of relief printing on paper, were developed in the 15th century in the Western world (although much earlier in the East). It has been suggested by archaeologists that a system of printing tiles was also developed. This would have simplified and sped up the making process. A wooden stamp would have been 'inked up' with a contrasting slip, then pressed gently onto the clay tile surface, thus transferring its colour. There are, however, serious practical drawbacks to printing wooden blocks 'inked up' with slip: getting an even covering of slip on the stamp, transferring colour smoothly and cleanly to the tile, and the problem of stamps clogging with slip. Anyone who has tried this suggested process will tell you that it is not possible to print like this. More likely is the scenario that a tile was coated with a contrasting slip, and when leatherhard was stamped with the wooden block. This would create an indented area, and the tile would later be scraped back in the usual manner to reveal the contrasting design.

Whatever the truth, the tiles do show widespread use of stamping and colour to create patterns and designs as only seen before with painted designs. The whole manufacture of these tiles in whatever form went out of fashion in the 16th century, and decorated tiles in Europe seem not to have reappeared until the end of the 17th century.

In the meantime printing on paper continued to develop, although at a slow pace by late 20th century standards. Primitive woodcuts on paper were simply rubbings from wood blocks, but by the middle of the 15th century prints from metal engraved plates were being made in roller presses, and the printing press had been invented.[1] Printing from engravings allowed much finer detail than block printed woodcuts, but their useful working life was shorter, as the

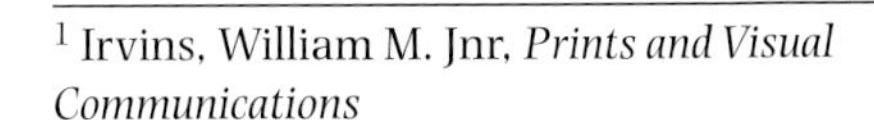

[1] Irvins, William M. Jnr, *Prints and Visual Communications*

Above: Original 13th century inlaid tiles in Winchester Cathedral. *Photograph by Diana Hall.*

Below: Replica inlaid tiles in Winchester Cathedral, made using traditional techniques, Diana Hall (UK) 1997. *Photograph by Diana Hall.*

Wood block printed tile, 1740. Cobalt oxide on tin glaze. *From the collection of the late Kenneth Beaulah.*

metal plates became worn. Still, developments in working techniques, and in the metals used gradually increased their life, and eventually the quality of image reproduction ensured that in the end, wood block prints were restricted to cheap illustrations. By the beginning of the 17th century engraving and etching was a finely developed art.

Although individual printmakers did explore the possibilities of the media (Durer and Rembrandt are well-known for having used etching as a creative medium), printing was, before the advent of photography, the main method of illustration and reproduction of images. As such, matters of taste, expression and creative possibilities were secondary to a system of illustration and reproduction in which economies and efficiencies were paramount. Skills of making plates, be they engravings in metal or wood, were divided. The original draughtsman or artist handed a work to a draughtsman for the engraver, who in turn passed work onto a specialist who made a preliminary etching on the plate or cut on the wood. Then the engraver worked up the plate, probably without ever having seen the original image for himself.

The very fact that engraving and etching were widespread meant that printing media, inks and oils must have been widely available. On paper the colour was probably a carbon lamp black or colours that would not survive a firing in a kiln. Potters at the same time were discovering metal oxides which would give increased ranges of ceramic colours, and these were painted onto tiles, plates and tableware. Cobalt blue first seen on Chinese ware became popular during the 17th

Tile made in Liverpool about 1760. Onglaze transfer print from engraving. *Collection of the late Kenneth Beaulah.*

century, on Dutch and English 'delftware', and colours of purple, yellow, green and orange were developed.

When eventually it was discovered that it was possible to use printed images to decorate pottery, the technology of the printer on paper combined with the knowledge of the potter.

Bernard Leach said that printing on pottery was 'a further stage in the division of labour, which, however necessary for mass production has destroyed the unity of conception and execution in the completed article'. But by this time the mass production of pottery had already ensured that the making of ceramic forms was divided into specific tasks, and the 'unity of conception' was already destroyed. The skilled decoration of pottery by hand painting gradually became a fringe decorative process, sometimes used to enhance basic printed images. Printing on pottery did mean the beginning of mass availability of decorated pottery wares.

The exact date of the discovery or invention of ceramic transfers is open to debate. Until recently, the most quoted is a John Sadler of Liverpool, who in 1749 is said to have observed children sticking bits of paper to broken pottery; six years later he swore an affidavit which claimed that he and a colleague had perfected a printing system. Robert Copeland in his *Spode's Willow Pattern* cites evidence that transfers were used in the Doccia factory as early as 1737, but whatever the ins and outs, by the end of the 18th century printing from copper plates was common in the pottery industry.

At first it was by a system of 'bat printing'. Copper plates were engraved by skilled craftsmen using a range of specialist tools. Images were built up using cut lines and punched dots. Then the plate was 'inked up' using a soft rag or cloth soaked in oil (one recipe instructs the use of '1 pint of linseed oil with a spoonful of pulverised umber' boiled for 30 to 40 minutes), and the surface wiped clean 'with the hand as in common copperplate printing'. The image on the plate in oil was transferred to the fired ceramic surface with the aid of a glue 'bat' approximately 7mm (0.25 inch) thick. This bat was pressed onto the copper plate, thus picking up the oil impression on the plate. Upon separation from the plate the bat was placed face up on a 'boss' (a soft leather bag filled with bran or wool) before being pressed onto the fired ceramic surface. The oil was transferred from the bat to the pottery surface which was then dusted with ceramic colour. The ceramic pigment stuck to the oiled area and so the image from the copper plate appeared. This method of transferring images was probably the first used, but because the amount of ceramic colour held by the oil after dusting was limited it was only suitable for on glaze, or enamel colours. Underglazes demanded a heavier deposition of ceramic pigment. Around this time, papermaking machinery was invented that made possible the production of large quantities of smooth, shiny paper. This and the development of printing presses that were operated by power and not hand had dramatic effects on the spread of information. They also made possible the development of a suitable tissue paper, which replaced the glue or gelatine bat as the main transfer medium for pottery printing. Harry Baker from Hanley made a patent in 1781 for four printing processes. What he described is a basic process for transfer printing:

> *Take a sheet of thin paper and having dissolved some gum arabic in water, spread it with a pencil on one side of the paper; let it dry; then take one part of balsam of amber and one part of Venice Turpentine for ink; put the copper plate on a charcoal fire or stove to warm and into this ink put such a colour as you please to, print with, and rub it into the plate . . . put the plate with the gummed paper through a rolling press, take it off the plate and put it onto the glass; rub it with a flannel to fix it to the glass, then soak in water and the whole impression will quit the paper and be left on the glass.*[2]

The use of paper transfers and the development of ceramic inks meant that underglaze colours could be used, and the process was adapted so that it was practical to use on biscuit ware. So the famous 'Staffordshire Blue' pottery evolved. Cobalt was the prime ceramic oxide used, and a wide range of blues were developed. Thousands and thousands of engraved designs graced the surface of Blue and White pottery during the 19th century. Whilst engravings were necessarily made to order for plates, cups, jugs and tiles, many were pseudo Chinese in origin, others were taken from engravings after classical paintings, or English old masters, or were specifically made for overseas markets (in particular North America). A number of particular styles were developed and an elaborate visual vocabulary of images and patterns built up. The basic process is still used some 200 years later at Spode in Stoke on Trent. A development in the 1950s called the 'Murray Curvex' system allowed for greater mechanisation of copper plate printing, but it is really an adaption of

---

[2] Copeland, Robert, *Spode's Willow Pattern and other designs after the Chinese.*

Above: Spode's Italian Blue design on bone china plate. 25cm dia. Based on book engravings showing exotic scenes fashionable in late 18th and 19th centuries. Central scene based on engravings after a watercolour by a Dutch artist. The border is an interpretation of Japanese Imari design. Still produced by Spode after 180 years.

Right: Detail of tissue printed from engraving for Spode's Italian Blue. Image is in reverse as it has not been applied to the ware. The image is slightly less precise in the finished article, because the cobalt 'bleeds' into the glaze, softening lines and producing varying intensities of colour.

Bowl, 38cm (15") dia., Charlotte Hodes (UK). Collaged slide on transfers and tissue copper plate engravings with painted onglaze colour. Painter Charlotte Hodes has worked at the Spode factory as an artist in residence on a number of occasions.

the bat print. A machine forces a solid convex pad of gelatine or silicon onto an inked copper plate engraving, and it then transfers the colour directly to the ceramic surface of the ware to be decorated. In this process, the copper plates are deeply engraved, and the inks used are in media made from synthetic oils. The process is usable only on 'flatware', but adaptable for onglaze and underglaze decoration.

Multicoloured prints from copper plate engravings were made from the 1840s. Early multicolour prints were from single engravings differently inked up, but five separate plates were engraved each one representing a colour in the design. Each separate colour was applied to the ware in the normal manner, allowing two days drying between each colour transfer. To enable accurate registration, small dots were incorporated into the designs. The results were beautiful elaborate images much used in their heyday on pot lids and containers for toiletries, meat and fish pastes and other delicacies. The process was labour intensive, and required above average skills, and in the

FR Pratt pot lid with polychrome print, *The Late Duke of Wellington*, circa 1860. *Photograph Paul Mason.*

Souvenir ware was produced with transfer decoration. This detail from a jug shows the *'Old Lane End Band at Practice'*. Earthenware, with pink underglaze print, 1837, manufacturer unknown. *Collection of Newcastle Borough Museum and Art Gallery, Newcastle Under Lyme (UK).*

economics of industry, multiplate engravings had a limited life.

The use of other colours for printing onto pottery is linked to differing processes of printing - relief printing, lithography and screen printing. Wooden blocks, long used in fabric printing, and prints from end grain of box wood, as commonly used in books at that time, had some life. In this process, instead of the incised area being inked up, the uncut surface was inked, and large areas of colour were printable. They were used on the same transfer paper process as engraving. The common problem in inking wood with ceramic colour was the amount of ink required to get a good line or even area of colour. Wood block prints are detectable by choking of fine lines with colour, or a darker line at the edge of block printed areas. Metal block plates (using zinc) were less problematic and have been used into the 20th century.

Woodblock printed tile, Maw and Co. Black underglaze. *Collection of the late Kenneth Beaulah.*

Sponge and rubber stamp decoration

The most successful 'block' or relief prints on a commercial basis were used on spongeware. The making of spongeware was a distinctively (but not exclusively) Scottish industry having its heyday in the latter half of the 19th and the early part of the 20th centuries. Early sponge decoration involved the use of the natural sponge in a loose and fluffy state, dipped in a medium containing ceramic colour, and pushed in contact with the biscuited surface of the pottery. The decoration was relatively random, and abstract. It was sometimes heavily loaded with cobalt to create 'flow blue' ware where the colour bled to such an extent that the white body appeared pale blue with the printed area showing as flowing darker blue patterns or shapes.

The first control of the sponge to print more regular decoration was achieved by tying the sponge with linen threads. This had the effect of pulling the sponge into a denser mass, the threaded areas determining the shape of the pattern.

To many people though, what is thought of as spongeware is the pottery decorated with the cut, smooth root of the natural sponge. The root of the natural sponge is dense but still flexible and absorbent, and its use enabled potters to make quite detailed stamps. These were used to decorate earthenware pottery with bright patterns and designs, and the ware was exported to the British Colonies, West Africa, North and South America. On some lines, sponge decoration was used in conjunction with copper engraved transfers but in general, compared to engraved transfer prints, the decoration was crude, and clay bodies used tended to be coarse. Reference to spongeware in books on pottery decora-

Sponge decorated plate produced by David Methven and sons of the Links Pottery, Kirkaldy, c. 1900. *Kirkaldy Museum and Art Gallery*

tion are few and far between, those that are made are generally rather dismissive, the phenomena seen as cheap and vulgar. It was perhaps viewed this way in its own day, for comparatively little (compared to transfer decorated and other contemporary pottery) seems to have survived. The Scottish spongeware industry folded in the late 1920s as a result of the Great Depression, although some ware was produced into the 1950s. In the contemporary pottery industry, spongeware, with its naivety, immediacy and 'hand made' quality, is making something of a comeback.

Associated with sponge printed decoration is the rubber stamp. Its first use was probably as a manufacturer's stamp for the bottom of ware. On copper engraved pieces the bottom stamp was included in the engraving, so the tissue print included it. It was simply cut out with the rest of the designs for the ware, and applied in the normal manner to the base of the pot. For sponge decorated pottery, the sponge was too crude to allow for bottom stamping, and so at some stage before 1890, the rubber stamp was introduced. The principle of use was sim-

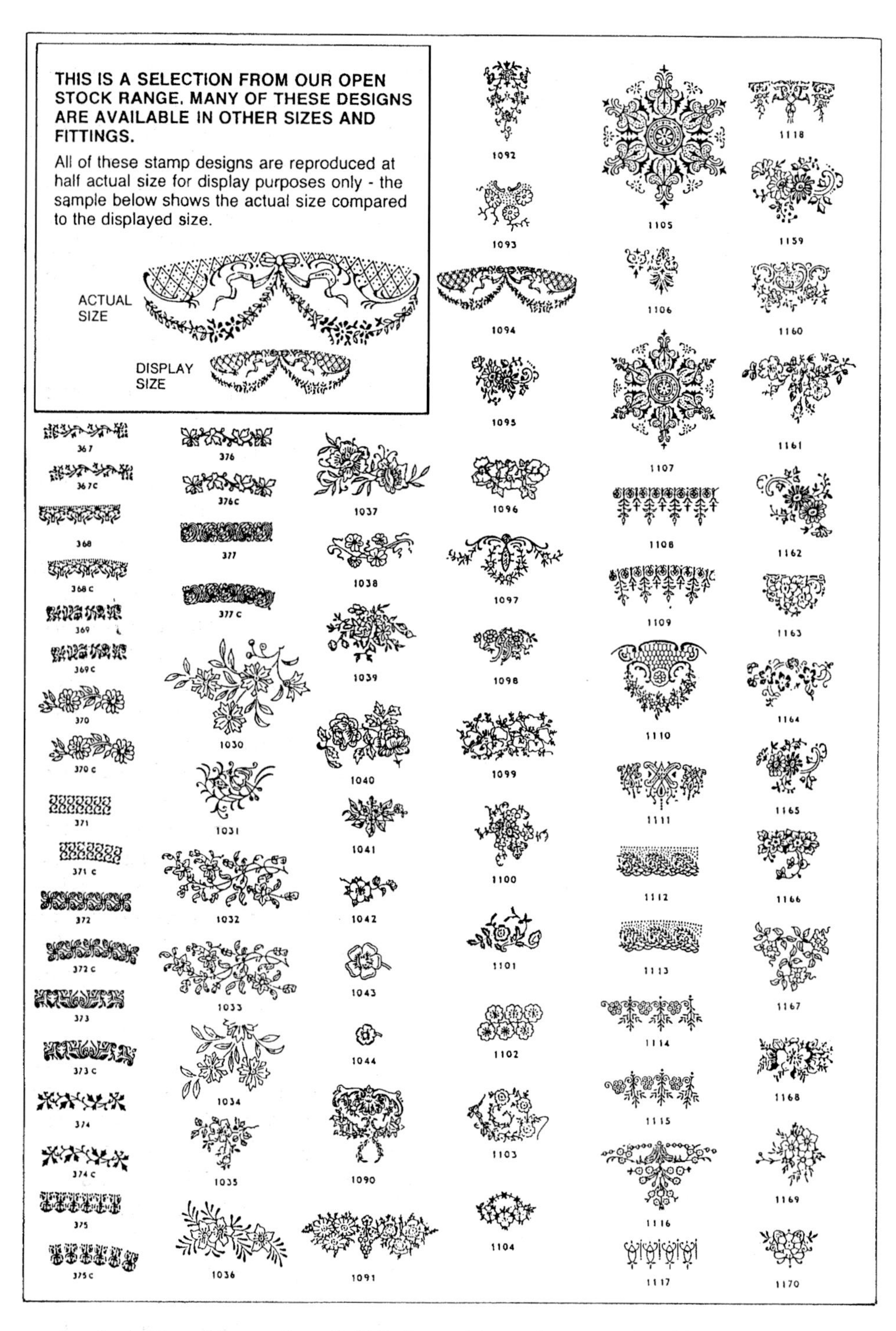

Small selection of images from K. H. Bailey and Son Ltd, open stock rubber stamp designs (reduced size).

ilar to the sponge stamp, but the rubber not being absorbent would only make one good print at a time.

Later use of rubber stamping came in the 20th century, when it was used to decorate gold lustred borders on bone china and porcelain. K.H. Bailey and Son Ltd of Stoke on Trent still have a large selection of 'open' designs of rubber stamps available today. Their use in industry has been on a steady decline as production has become increasingly automated, but like sponging, its 'hand made' quality ensures its survival for the present.

Three processes have revolutionised printing since the invention of engraving: Lithography, Photography, and Screen Printing (or Serigraphy – drawing on silk).

Lithography

Munich playwright/composer Senefelder discovered the process of lithography in 1797. Lithography (from the Greek 'lithos'- 'stone' and 'graphio' – 'draw') relies on the natural repulsion between grease and water. In its most basic form, an image is drawn on a prepared slab of limestone with a greasy crayon. The slab is then flooded with water, and then inked up with an oil-based ink. The ink is repelled by the water, but the greasy areas pick up the ink. Paper applied to the inked up slabs picks up the image from the drawn area. By repeating the process, numerous identical images can be printed. At first lithography was exploited by the paper printer, and it was in continental Europe that it was adapted for use on pottery. It was some time before the British pottery industry took it up with any enthusiasm.

By drawing separate blocks for separate colours as with the copper engravings, multicolour ceramic prints were developed. The process of making the plates was less labour intensive, the quality of the image was different, more like original paintings or drawings. Specialist adaptions of tissue paper were produced to make the process workable. Duplex paper (used until relatively recently) consisted of a thin sheet of tissue with a much thicker paper backing. The thick paper backing allowed for stability in printing colours in register, and this was removed before the image was transferred to the ware. The tissue paper possessed the flexibility for transfer to curved ceramic shapes. Ceramic surfaces were painted with a weak varnish, which when tacky held the lithographically produced print. As with the copper plate transfers, the backing tissue paper was soaked and rubbed off before firing.

Zinc plates have since replaced stone. The porous nature of the stone is simulated by graining a slightly textured surface made up of tiny recessions capable of retaining moisture or grease.

The lithographic process was, and is still used, primarily for onglaze or enamel colours. The general consensus has been that as with the bat printing process, ink thickness is fine and unsuited for underglaze decoration. Still, after the introduction of photographic lithographs, it was possible to photographically transfer images from engravings and wood blocks onto lithographic plates, and it has been suggested that a number of printed tiles and underglaze decorations on pottery towards the end of the 19th century were lithographically produced.

Screen printing

Screen printing, a development from stencilling, first appeared in Japan and its invention is attributed to Yutensai

Miyassak in early 18th century. The problem with stencils is that isolated parts in an overall design need to be attached to the main, thus

P becomes P

Bridging sections became very obvious as designs become more elaborate. One solution developed by the Japanese was to cut two stencils exactly the same shape and size. Human hairs were glued across the shapes of the first stencil, holding them all in position, and the second stencil was glued on top of the first, sandwiching the hair between them.

Decoration by painting or dabbing colour through the stencils was then possible, and the resulting image did not show revealing bridging struts as these were the hairs. The story goes that the hair support was eventually replaced by silk mesh.

Screen printed tile made by Carters of Poole, Onglaze colours, made about 1952. *Collection of the late Kenneth Beaulah.*

It wasn't until the early 1900s that screen printing, as we recognise it today, was in use. Its first industrial application was onto fabrics, and by using a cut stencil process. In 1939 in New York Anthony Velouis developed a process which dispensed with the cut stencil. The blank screen was drawn onto with a waxy or lithographic crayon, then washed with a glue solution. When the glue set a solvent was poured over the front and back, dissolving the wax; creating open areas of the screen where the image was originally drawn. Subsequently, photographically sensitive emulsions made screen printing an even more versatile medium.

In ceramics, the pottery industry first used screen printing for tile decoration but not until the 1950s in the UK. The first silk screen printed tiles in England were attributed to Carters of Poole. Now it is the prime process for ceramic decoration in industry, either in the printing of decals or direct printing. The process is so versatile that not only does it allow printing on the flat, but it is possible to print directly onto cylindrical shapes (mugs, etc.), and to print with wax resist, glaze, onglaze, underglaze, heated thermoplastic colours (so that a decorated piece does not need another firing).

Its combination with photographic processes was the key which unlocked the door to its mass use in industry. The other key was the development of 'decals': Ceramic colour is printed onto a special gummed paper, over which a thermoplastic layer is printed in liquid form. The 'covercoat' as it is called, sets on drying, including the ceramic image in its body. When placed in warm water it is possible to slide off the covercoat sheet with the ceramic image in place. The residues of the gum allow it to stick onto a ceramic surface. On firing the plastic burns off leaving the printed image in or on the ceramic surface.

Photography

The first person to make permanently fixed images with a camera was Joseph Nicephore Niepce. In 1827, using a light sensitive coating on a zinc plate, he exposed it through a translucent original engraving. The zinc plate was subsequently etched and several prints made.

The first printing of a photographic image on a ceramic surface was undertaken by Lafon de Camarsac in France in 1854, using a gum bichromate system. This relied on the light sensitive chemical potassium bichromate. Mixed with gum arabic, and a sticky substance like honey, it was coated onto a ceramic surface and exposed to light through a positive transparency. Where light hit the surface of the tile, it fixed the gum; where it was darker the substance, remained proportionately sticky to the amount of light falling onto its surface. The result was a latent photograph, revealed when ceramic pigment was dusted onto its surface. Where the gum has hardened no pigment adhered, but pigment did adhere to those areas un- or under-exposed. It appears that he also used a system where ceramic pigment was mixed into the light sensitive emulsion. In 1868 Lafon de Camarsac was marketing a system of producing portraits on porcelain, and although a number of companies were said to have taken up the system, it was one requiring a degree of skill to work effectively and consistently, and wasn't one for mass industrial use.

Porcelain photograph on gravestone in Aosta Valley Italy. *Photograph by Amanda Spencer-Cooke.*

Still a fashion for ceramic photograph on plates did develop, and perhaps the most interesting use of the technology was in the production of ceramic photographs for gravestones. In some traditions it became customary to place photographs of the deceased on their graves. Unfortunately, paper photos do not stand up well to the ravages of time and the elements, but ceramic images, captured under a glaze, do. In areas of France, Italy and Switzerland the practice is still followed, ceramic photographs are common sights in graveyards, and it is possible to order them from specialist manufacturers.

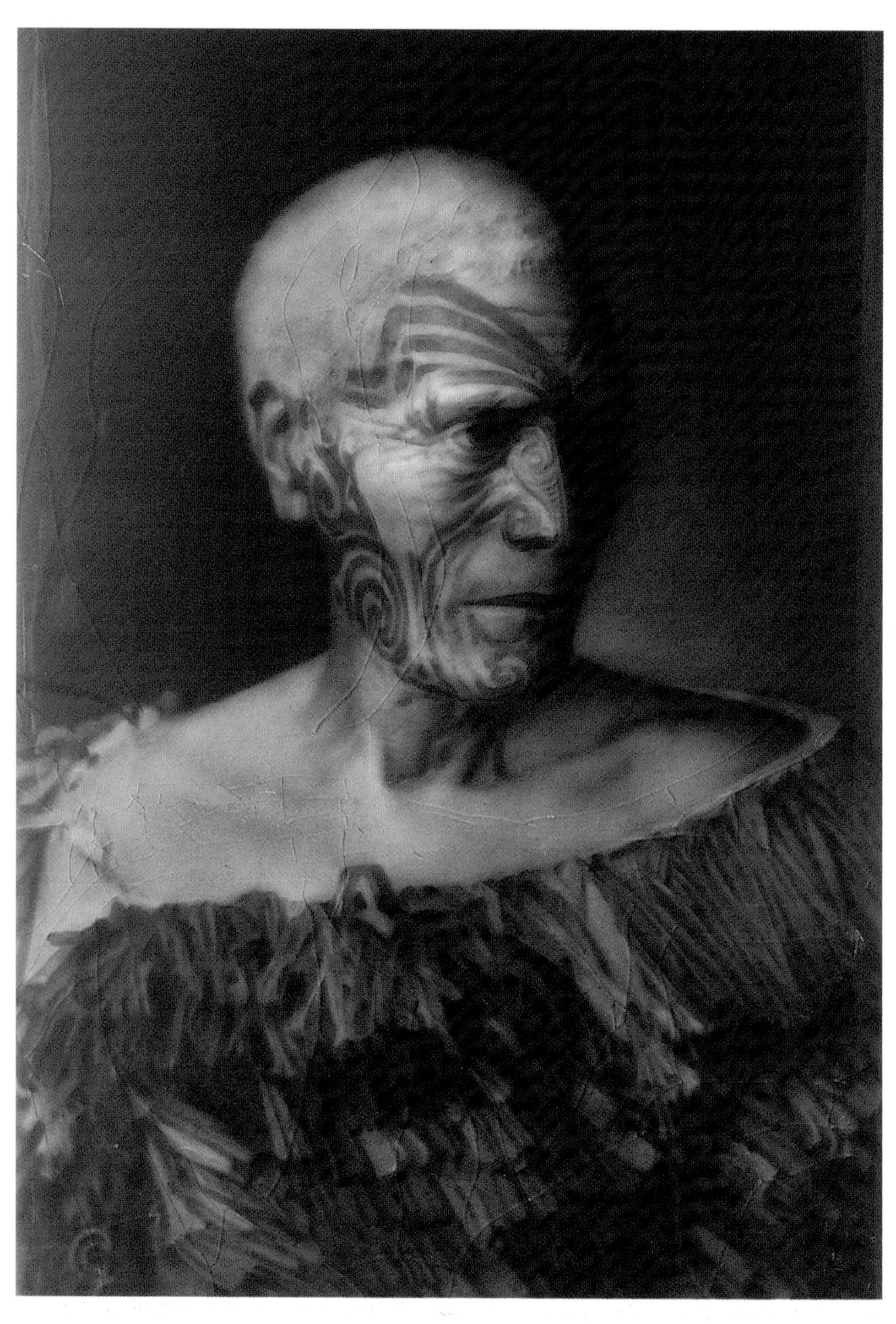

Tile with portrait of a Maori, Ivari Netana, by George Cartlidge for Sherwin and Cotton, c. 1897–1924. *City Museum and Art Gallery, Stoke on Trent.*

In the fevered clamour for new inventions that characterised the last half of the 19th century, a number of other photographic processes were developed, and some were tried out in the ceramics industry. One involved a bichromate process with gelatine in the photographic production of relief moulds. Tiles made from the moulds were glazed with a tinted, transparent glaze. Where the glaze was deepest, the colour was darkest; where the glaze was thin, the white of the tile showed through.[3]

However, photography's greatest impact in printing and ceramics lies not in direct photographic emulsions but in the use of photosensitive emulsions used in intaglio, lithography and screen printing.

Current industrial uses and developments

Screen printing and lithography now dominate the ceramics printing industry. Synthetic media based on hydrocarbon solvents have made many things possible. High tech machinery and equipment are capable of huge production runs, with exacting colour matches. New coloured inks make any colour that is printable on paper, printable on ceramic. In response to the health and safety and environmental problems of solvent-based products, new ultraviolet (UV) curing media are being researched. The old glue bat process of transferring colour and designs to ware has been resurrected and refined. In pad printing, a process that involves screen printing, and offsetting, gelatine or silicone pads now transfer images that have been screen printed onto a flat plastic or glass surface, onto a curved ceramic one.

The advent of microprocessors and the dawning of the Computer Age has resulted in new developments industrially, many to do with mechanisation, but also to do with image production. Much of the development appears to be heavily based on investment in machinery, hardware and software. How much of it is of use to the small-scale producer, or the artist/ceramist printmaker remains to be seen.

Ironically, at a time of technological advancement, old processes are making something of a comeback, in particular, sponge printing is enjoying a new lease of life. It is to some extent 'fashion' led, but it is important to remember that spongeware production ceased in the Great Depression, due to the stringent economic times, not due to its lack of popularity. Today, several companies are producing spongeware on a reasonable scale, as before much for export. The advantage that a simple non-machine process has is that initial setting up costs are relatively small, and the development of new lines and products does not rely on huge investments. Thus spongeware has developed from its traditions, being innovative and still marketable, whereas much of the transfer decorated commercial ware is either in a time warp, producing work from engravings or designs made last century or safe mass marketable decorated wares. There are exceptions, but these are few and far between.

Studio practice

Although artists began to explore the creative potential of printing on paper, on a commercial and creative basis, from

[3] Recent research at the Fine Print Research Department, University of West of England has identified methodologies for reproducing this technique.

relatively early days, nothing of this nature[4] developed on the ceramic surface until Sam Haile's experimentation with screen printing on clay at Alfred State University, USA in the early 1940s[5], and Picasso's exploration of the potential for printing lino cuts in clay at Vallauris in the 1950s.[6]

The reasons for this delay in experimenting with the ceramic are partly technological, but they are also to do with the study and classification of the visual arts; 'In terms of hegemony of art forms, certain are privileged over others'[7], and the effect of the early exclusion of ceramic practices and traditions from the definition of 'the fine arts' devalued the practice of painting and printing on ceramics. The study of (ceramic) practice and methodology was not taught in the academies, and the development of skills in ceramic printmaking were severely restricted by access to the necessary materials, knowledge and technology. Until the late 20th century, these have effectively developed alongside and inside industrial production.

As well as historical factors discouraging artist/printmaker exploration of the ceramic, the growth of the studio pottery movement in the UK in the second half of the 20th century put a further damper on the potential for exploration of printed surfaces from the clay side. The identification of the 'artist potter', and the study of ceramics that evolved in Britain in particular originated in the philosophies and writing of Bernard Leach, and the Design movement linked to industry. The guiding philosophies behind both have been concerned with lifestyle, form, production and design. For them, marks on ceramic surfaces are for 'decoration' and entirely subservient to form and function.

In the USA, although the Leach influence was strong, other forces were also at work, and in the 1950s and 1960s in a culture unfettered by centuries of tradition and academic practice, Abstract Expressionism, Pop Art, and other movements quite outside studio ceramics transformed artistic practice in the medium. Ceramic movements, including Funk[8] and the Super-Object[9] spawned a much greater diversity of work in clay. Although slower in Europe, recent years have seen a radical change sweeping through the bastions of studio pottery and an openness to change. An increasing exposure to ceramics from around the world, through international magazines like *Ceramics Art and Perception*, and the increasing use of clay as a medium for sculptors, painters and contemporary artists has opened the field in a way unthinkable just a few years ago.

There is increasing recognition of the traditions of print in ceramics, and ceramics in print. Contemporary practice is now diverse, and interdisciplinary. The blurring of the lines between 'artist' and

---

[4] Artists have been employed by the ceramics industry for centuries, but generally to design for production (e.g. Eric Ravilious at Wedgwood, Scottie Wilson for Royal Worcester), see chapter 3 *Painted Clay, graphic arts and the ceramic surface*, Paul Scott (A&C Black 2000).

[5] See *Sam Haile, Potter and Painter*, Bellew 1993.

[6] See *Vallauris Céramiques de peintres et de sculpteurs*, Réuniondes musées nationaux/ Museée de Céramiqu et d'Art moderne De Vallauris , Paris 1995.

[7] Conrad Atkinson in *The Plate Show* catalogue, Collins Gallery Glasgow 1999.

[8] Exemplified by Robert Arneson at UC Davis.

[9] Robert Shaw, Victor Spinski, surreal *trompe l'oeil* porcelain assemblages.

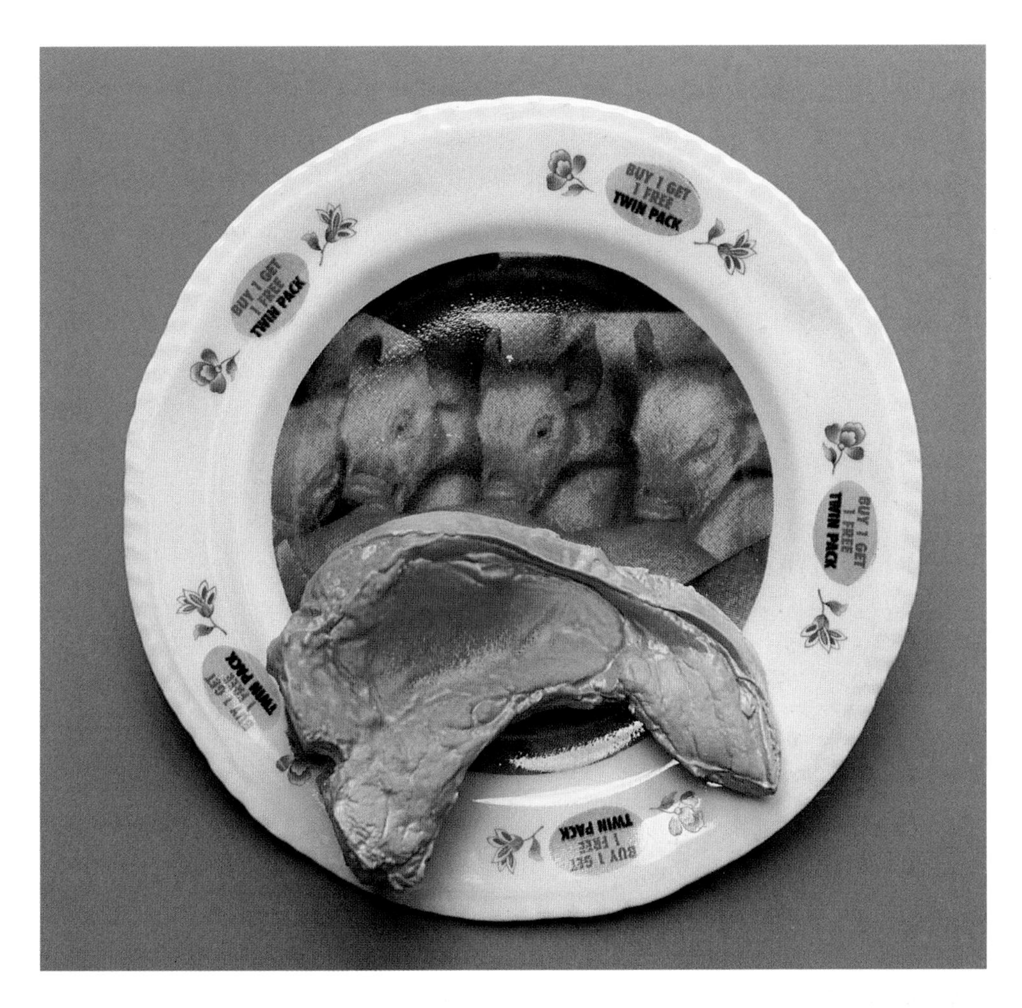

'craftsperson' gathered pace in the early 1960s and 1970s and many have gone through a system of further education where the demarcation between differing art forms and disciplines has been eroding. Many of the artists featured in this book are not welded to the ceramic as their main medium, but are variously painters, printmakers and other contemporary practitioners. Others now specialising in ceramics have come from other disciplines, transferring and adapting their skills and techniques to a different medium, and in doing so have created new methodologies and material connections.

B.O.G.O.F. (series) 2001, Caroline Taylor (UK). Slip-cast plate and pork chop, 24cm dia., with four colour, screen printed transfers and commercial transfers.

Chapter Two
Relief printing

The strict dictionary definition of print is 'to impress'. This means that many ceramic surfaces are 'printed', from the thumb print of the thrower or hand-builder, to the stamped bases of pots. Some individuals exploit clay's ability to receive a relief surface, faithfully recording the imprint of the stamp to produce complex and detailed decoration. Steve Scholefield (New Zealand) is one such, impressing the clay surface of his forms with small, plaster and biscuit-fired clay

Above: Stamp near the base of a wood-fired bowl, by Alastair Hardie (UK). *Photograph by Andrew Morris.*

Right Plate, 35cm (14") dia. 1250°C. Steve Scholefield (New Zealand). Stamped with carved plaster stamps, each square with a pattern represents a single stamping.

Detail of a salt glazed goblet by Stuart Whately, Edinbane Pottery, Isle of Skye. Thrown pots are rolled on bisque or plaster stamps producing the relief print of the fishes. The salt glaze brings out the line and form of the fish shapes.

stamps, later enhancing the 'printed' surface with a stoneware colours and firing.

For the printer, relief printing (block printing) is printing from a design, standing in relief. Raised areas generally receive a coating of ink from a roller, or in the case of rubber/sponge stamps from colour on a pad or tile. Printing takes place when the inked surface comes in contact with paper or ceramic surface.

Using industrial relief processes

The printing industry, in particular newspaper production, used to rely on a system of relief printing from lead type. Many potters will have used lead stamps to print names under pots, but others have been more adventurous with the technology. Paul Mason discovered newspaper 'flongs' some years ago. Flongs are cardboard sheets impressed with lead type. A page layout was made up in lead type, which was then printed into the surface of a thick, soft dampened sheet of card. The type created an embossed surface which was subsequently used to cast a lead 'positive' sheet from which printing could take place. It was used when it was necessary to send typeset pages long distances. This cardboard flong was lighter – and therefore easier and cheaper to handle than a bed of lead type.

Having acquired a number of flongs, Mason saw the possibilities of using these with clay. In order to make a printing plate, it was necessary to make a cast from the flongs. They were stuck to a rigid surface, and lightly coated with polyurethane varnish to protect the paper-like surface. Instead of using lead he used plaster of Paris, cast about 1in. thick. The resulting plaster cast was a relief block of the paper flong from which ceramic prints could be made. Some were slip cast from the plaster, in other pieces he used the plaster block to impress into clay slabs.

In a similar way, Lenny Goldenberg has combined an interest in clay with an interest in writing, producing a series of ceramic books imprinted with actual lead

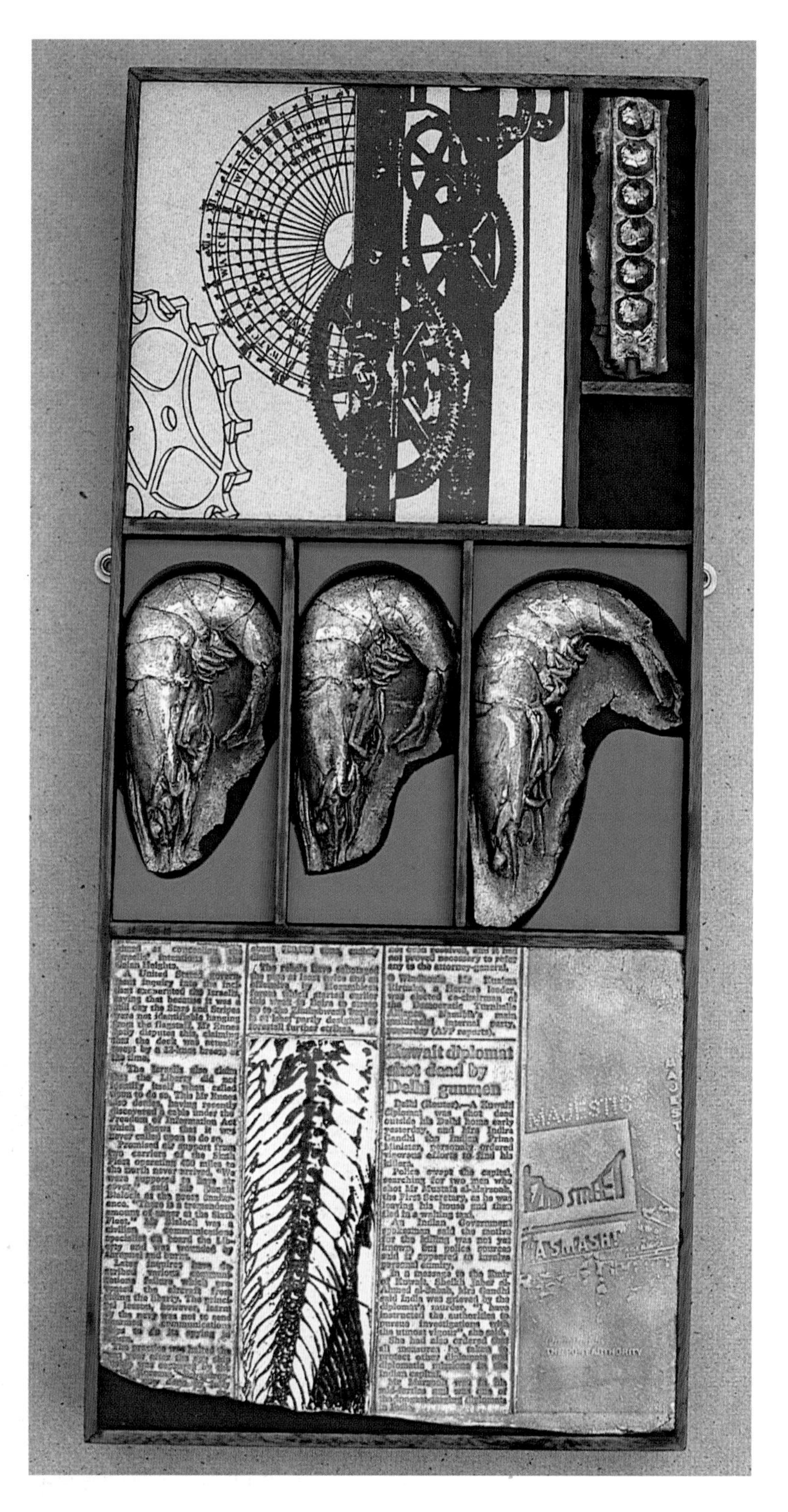

Prawn Clock Box from *Hot off the Press* Exhibition, Paul Mason (UK) 1995. Newspaper flong print, pressmoulded prawns, silk screen printed underglaze decals in stained beech frame, 50cm x 23cm (20" x 9").

type. Typesetting is a slow and laborious process, and he has discovered a simpler, quicker way of getting relief type for printing with. He uses a computer to make up a page layout, and sends the resulting print out to a local printer who makes up a relief plate from a flexo, or, polymer plate. Made up of a flexible polyester or thin steel support sheet with a light-sensitive polymer layer, a relief plate is made up from artwork on a film negative. After exposure, it is washed in warm water where the unexposed surface washes away, leaving a relief surface, suitable for printing from[1] It is a process still in use today, mainly for the production of cheap printed cartons.

The plate is flexible, and Lenny uses

[1] The process is basically simple, but specialist equipment includes a UV light source with a vacuum or, at least, some way of getting secure contact between the film artwork and the plate.

Above: *The Cello Suites: Erotic Misadventures, the Breast Book,* Lenny Goldenberg (Denmark), approximately 30 cm in length. Hand-built and pressmoulded ceramic book. *Photograph by Vita Lund.*

Below Lenny Goldenberg (Denmark) pressing curved composing stick with lead type into clay book. He rolls type from left to right across the clay page. *Photograph by Vita Lund.*

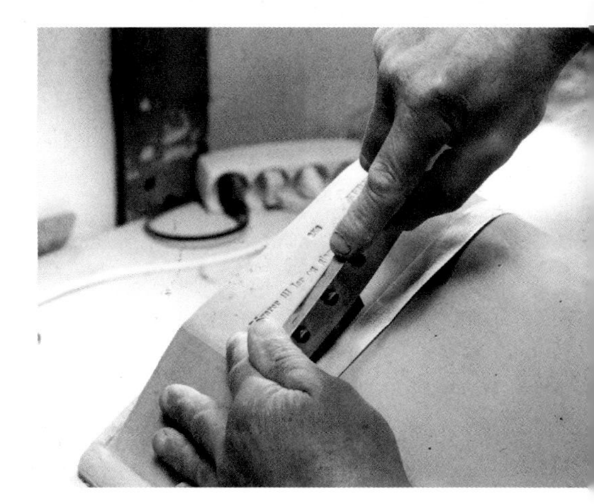

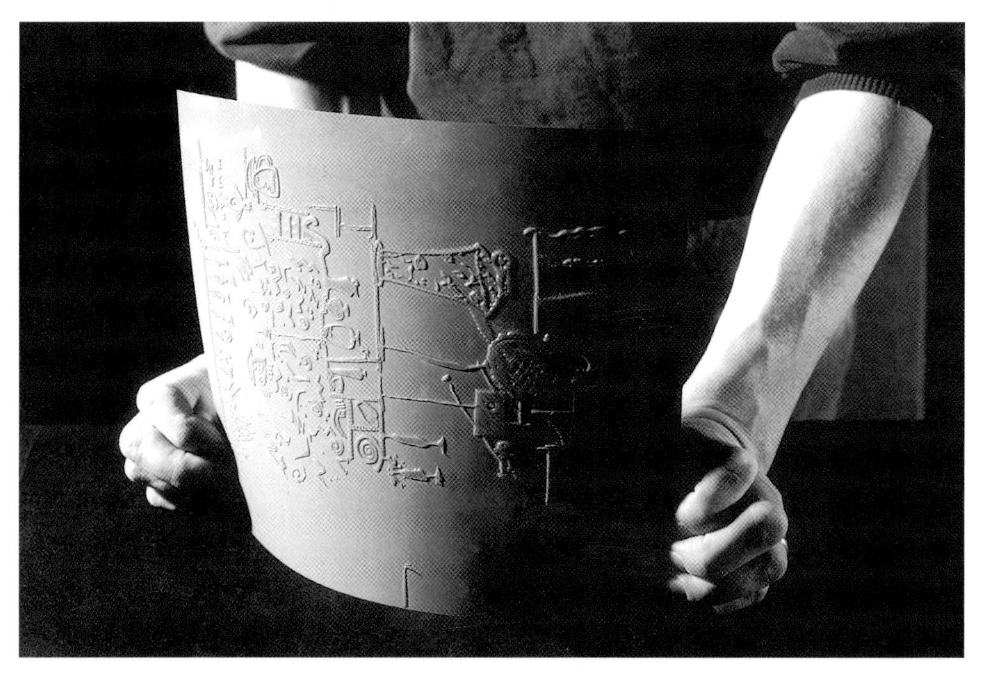

his by rolling it into a leatherhard clay, impressing the typeface into the surface. After biscuit firing, ceramic pigments are painted into the indented surface, and the excess wiped away, revealing the typeface. He has observed that photographs produced in this way do not make a deep impression and are not as effective as the typeface. Recent work by Helen Smith at the University of the West of England has further explored the potential of these flexoplates for use with

Top: Design on flexoplate. This can be used to impress clay, or be inked up and printed. *Photograph by Andrew Morris.*
Above, left: *Felixstowe Beach Signal box.* Porcelain paperclay blind embossed with photographically etched flexoplate displayed on light box. Funded by the Arts and Humanities Research Board, Smith has been researching the potential of printmaking in lithopanes ('porcelain transparencies').
Above, right: Flexoplate of *Felixstowe Beach Signal box*. Helen Smith University of the West of England.

ceramic surfaces. She has observed that tone and three dimensionality are opposite ends of the equation, with more tonal detail meaning less three dimensionality to the plate and has been experimenting with images on the computer using Adobe Photoshop to control the tone/3d balance.[2]

It is also possible to ink up flexoplates with ceramic ink, specially prepared with copperplate oil, or other similar media.[3] A print transferring colour can be effected by placing the inked up plate onto leatherhard clay, and rolling the back of it with a rolling pin, or by transfer printing onto fired clay surfaces using a pottery tissue transfer.

[2] See Chapter 5 for further details of computer use.

[3] Intaglio prints can also be made with this material, see Chapter 3 Intaglio printing.

Relief printing of the artist-printmaker

To the artist-printmaker, relief printing means linocuts and woodcuts. These are made by cutting away those areas that are not to be printed. In the case of woodcuts, the work is done on a plank of wood cut along the grain. The presence of grain has an influence on the cutting edge of the tools, and this is often clear in a finished woodcut print. Lino on the other hand has no 'grain' to it, and can be cut in any direction without interference. When using lino, it is possible to soften it by gently warming it before use. This makes for smoother, easier cutting. A smooth, leatherhard clay slab (porcelain, or ungrogged earthenware) can

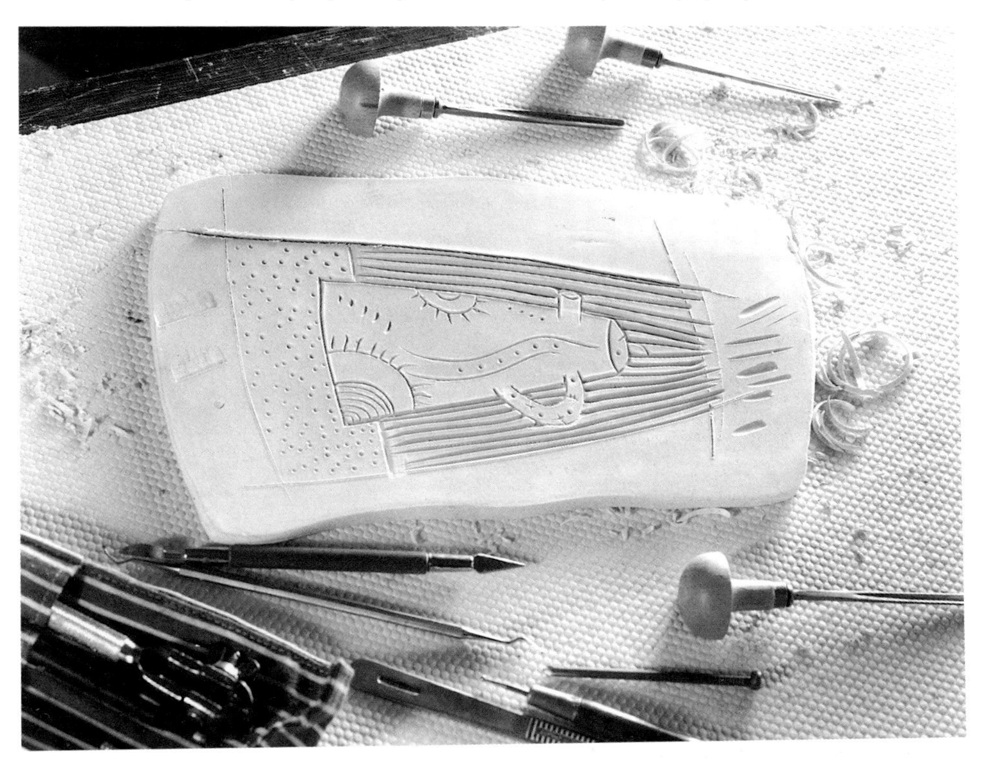

Porcelain slab cut as a linocut, ready for trimming. *Photograph by Andrew Morris.*

provide a useful testing surface for those new to lino cutting. The surface requires much less effort to cut, but it does help to have a familiarity with the linocutting tools. It is even possible to print from the leatherhard 'clay cut', and of course, the tile can be fired in a conventional manner.

In spite of the popularity of wood engraving (done with specialist tools on end grain of box or holly), woodcuts, and lino printing with fine art-printmakers, these disciplines are not well-represented in the ceramics field. The wood block ceased to be of importance in industry

Left: Print from porcelain 'lino cut', Paul Scott 1993.
Below: *Fish tray* Jacqueline Norris (UK) 1993. T-material, with lino printed image.

Jacqueline Norris removing uninked linocuts from a slab of clay. The 'prints' were made by passing a slab of clay and linocuts through a clay slab roller.

last century, in spite of the fine density of colour and character of line that these produced.

In recent years a number of ceramists and printmakers have begun to experiment with lino printing on clay. Lino cuts printed on clay not only convey colour to ceramic surfaces, but also create 'embossed prints' as the flexibility and depth of soft or leatherhard clay

Juliette Goddard (UK) Jug with lino print.

Flower Checkerboard, linoprint blue/white stoneware 15cm (6") square, Carol Wheeler (UK).

allow a full impression of the whole linoblock, cuts and all.

Juliette Goddard cuts designs into lino, paying as much attention to the lines created in the 'cut out' areas as to the areas left in relief. Ceramic inks are made by thoroughly mixing ceramic pigments with copperplate oil on a glass slab with a palette knife. A sticky but smooth consistency is necessary for the ink to work. This is rolled onto a glass slab with a printing roller, which is then used to transfer the colour to the linocut. The inked linocut is then printed onto clay slabs which can be used for slab building.

Richard Slee has also used lino prints to create pattern ceramic forms. His

Left: Dirk Hagner (USA), clay lino print. Richard White (USA) uses linocuts to press into porcelain slabs, but has developed a technique similar to the manufacture of medieval floor tiles. The lino has a tendency to stick to the damp clay so White recommends prior spray with WD40, olive oil or dry dusted clay to help release the block. After printing the clay is painted with a contrasting slip and left to dry. When sanded gently with steel wool it reveals as much of the image as choosen. 'The results are as perfect and detailed as a print on paper, and of course you can bend it as you wish since it is clay. If you use a slip lighter than the clay you start with, the result is a negative.'

Anvil Plate Richard Slee (UK) 1988, 30 cm (12") diameter. Pressmoulded, with lino print and rubber stamped border, 1080°C.

approach has been nearer to the artist-printmaker in the quality of image produced, but his pieces are very much within a ceramic tradition. Using fat oil with a little turpentine as the medium into which he mixes ceramic pigments, he prints from a linocut onto pottery tissue, which is then applied to the biscuit surface of forms whilst the ink is still tacky. Ninety per cent of the printed image transfers from the tissue to the bisque surface, and whilst he does remove the pottery tissue before firing the work, he thinks this is probably unnecessary.

Relief printing traditionally associated with ceramics

Rubber stamps

Rubber stamps can be made to order, from any design or drawing. They can be quite complex and individually decorative. K.H. Bailey and Sons Ltd of Stoke on Trent still stock a large range of 'open' designs which may be purchased at a reasonable price. Rubber stamps are not usually used for large designs, for in general, after about an inch square, the larger the stamp, the more difficult it is to get a good print.

It is also very simple to make your own rubber stamps using cut up soft pencil erasers (Staedtler, or W. H. Smith pencil

For ilya, Conrad Atkinson(UK/USA) 1996. Earthenware plate, 25cm (10") dia. with underglaze stamps and written lustre legend.

eraser EAN13840). The simplest can be made using a scalpel-type of knife, cutting the eraser into shapes desired for printing from. More elaborate designs can be worked upon using a few simply made tools. George Thompson in his book *Rubber Stamps and how to make them* suggests making tools from 'one fine needle point, and a thick darning needle sharpened to a chisel point on a carborundum stone'. These can be turned into tools by pushing them into the end of old paintbrush handles, or even sticks from the garden.

Working with the rubber is mainly a matter of common sense. It is probably better to draw complex designs onto the rubber first in black ink with a felt pen or Indian ink and a paintbrush. When cutting the design, avoid undercutting the printing surface as this produces poor prints and can cause the stamp surface to break off altogether. Stamps can be used as they are, but they are easier to use with some sort of handle made from dowelling, cork or wood off cuts stuck to the back of the stamp. If using handles, ensure that the seals do not overlap the

Collection of John Calver's stamps and rollers

handle by more than 3 or 4 mm (1/8").

When stamping designs on paper, it is common to pick ink up from a soft pad soaked with pigment. Commercially for ceramic use, ceramic pigment is mixed with an oil-based media (fat oil, copper-plate oil) with a palette knife, then rolled out onto a gelatine pad. The rubber stamp is pushed onto the gelatine where it picks up the colour, and then it is stamped onto the ceramic surface, depositing the pigment. An inked piece of glass will work quite adequately. Water - based media are not as successful as oil-based ones, but they will work.

If printing onto bisque with oil-based media, the work will require a firing before applying glaze so as to burn off the oil. If pieces to be printed are strong enough, then printing onto leatherhard clay will avoid the extra firing.

Sponge printing

Sponge printing has a long ceramic tradition. Unlike rubber, sponge stamps are porous and hold a reservoir of colour. They can be used for much larger designs, and repetition printing is easier because the sponge will hold colour for a number of prints before needing to be re-inked. Sponges are coarser and, in general, will not give the fineness of image that a rubber stamp can. In the early Scottish sponge decorating industry, it was the dense root of the natural sponge that was used for making the sponge stamps. Today, modern industry produces many different types of foam, most of which can be used for sponge printing.

Sponge stamps are usually used with water-based media. A number of pottery companies now produce spongeware on an industrial scale again. The method has also found favour with individual ceramists, who have found its qualities and ease of use seductive. For general use, ceramic pigment mixed with a little underglaze media, or even just water, and painted onto a tile or glass plate will provide a perfect reservoir of ink for use.[4]

John Calver uses sponges dipped in

[4] Specially formulated *Potters Pads* are now available with a range of colours ready for stamping.

Sponge decorated plates awaiting glazing at the Nicholas Mosse Pottery in Ireland. Although each plate is printed with similar stamps and to a basic design, each one is slightly different to the other. This gives the feel of originality, and the 'hand made' touch to the work of an industrially produced object.

Above: Contemporary sponge decorated mugs and bowls awaiting packaging at the Nicholas Mosse Pottery, in Ireland.

Right: John Calver sponge decorating platter.

bowls of stained slips to decorate leather-hard work. He also impresses using clay stamps, trailed and poured slips, and later, overlapping glazes to produce the rich surface qualities on his range of domestic ware.

Veronica Newman uses sponged designs on biscuited porcelain, subsequently glazing with opaque glazes, and Bronwyn Williams Ellis uses them to decorate with underglazes on top of tin-glazed commercial tiles before refiring so that the colours sink into the glazed surface. All sponges can be cut with sharp scalpel-type knives, but a number of ceramists also use hot cutting. John Calver uses a hot wire, but he advises that such activity should take place outdoors or with good exhaust ventilation because of the noxious fumes given off by burning foam.

Yorkshire Weather, Jim Robison, (UK/USA) vase form 50cm (20") tall. Cloth stencil with porcelain slips brushed over the surface. Pasta roller used to impress linear details into clay.

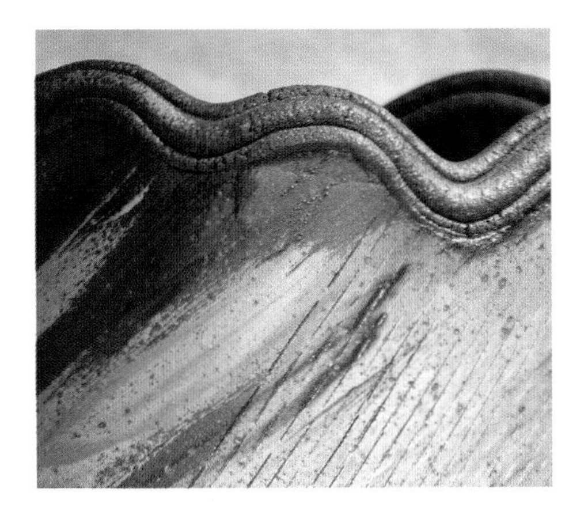

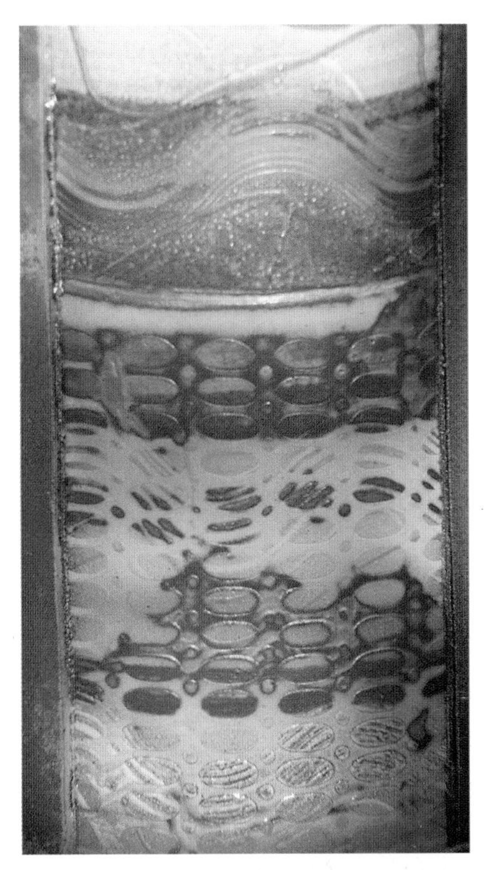

Above: Detail *Yorkshire Weather* Jim Robison (UK/USA).

Right,: Detail *Landscape vase,* Jim Robison (UK/USA). Cloth stencil was used several times and reversed. The stencil produces a coloured print itself while the contrasting slip is brushed through the openings.

Below: Cloth stencil(originally a curtain fabric) is rolled into the clay slab. Slips are applied with brushes and rollers. Pieces are constructed when slabs have dried to a leather hard condition. Jim Robison (UK/USA).

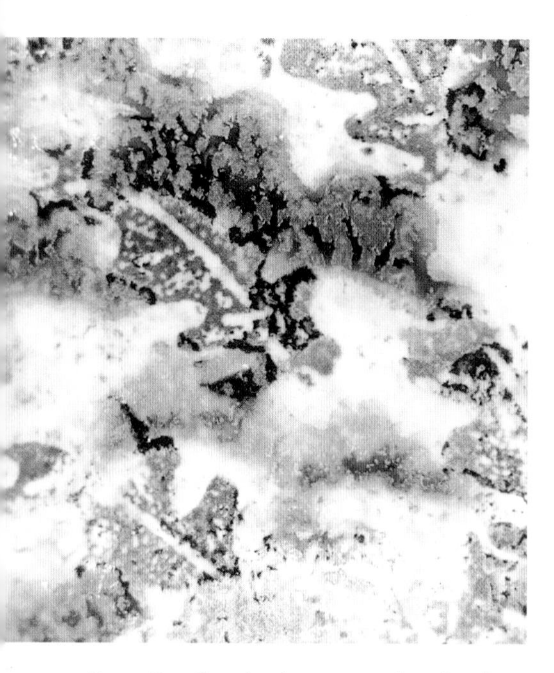

Above: Detail underglaze sponged surface by Veronica Newman (UK)

Below *Chinese dragons*, Bronwyn Williams Ellis (UK), sponge printed oxides/carbonates on tin glazed commercial tiles.

For many years John Pollex was a respected maker of traditional slipware, but in the mid-1980s he became heavily influenced by painters Howard Hodgkin, Robert Natkin, Ben Nicholson, Hans Hoffman and David Hockney. These influences clearly show in his work which has painterly qualities quite unlike any other spongeware. It is true that most of his surfaces are produced by a combination of painted and sponged marks of intensely coloured slips, but many of the seemingly 'painted' qualities of the work are applied with sponge stamps.

John works on thrown and altered shapes, using commercial body stains mixed with white slip. The colours are applied to leatherhard work which has previously been covered with a black slip. Some of the sponge stamps he uses are industrial offcuts or byproducts, producing grids and formal shapes. He fires to earthenware temperatures which ensures a wide range of bright colours.

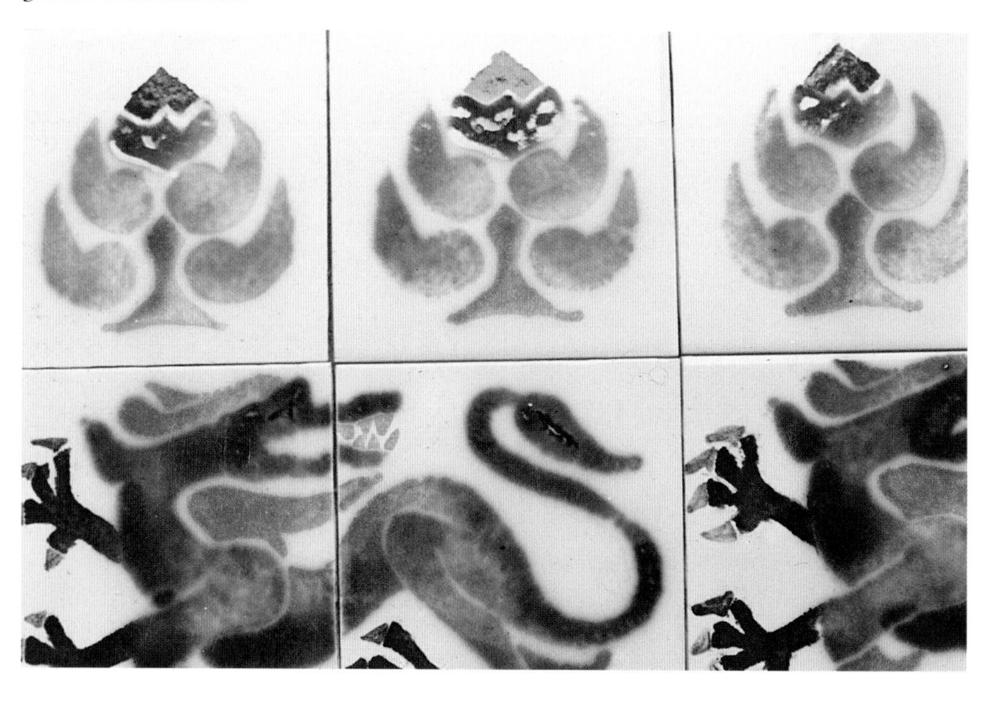

Sponging and stamping with lustres

A number of ceramists have adapted traditional sponge printing techniques for use with precious metal lustres. The resulting work is quite different in character from what is thought of as sponge-decorated ceramics. This is due in part to the surface quality that these materials impart to an object. Lustres are very thin layers of precious metals applied to prefired (usually burnished or glazed) ceramic surfaces. Being precious metals, printing with large, conventional stamps could prove an expensive business so they tend to be used with economy, and on a smaller scale.

Plate 32cm (12.5") dia., John Pollex (UK). Sponge printed and painted slip.

There are ranges of coloured lustres as well as purely metallic finishes. Colours are translucent and give iridescent qualities to a surface, while metallic lustres are more opaque. They are usually supplied in liquid form ready for painting. The metallic and other colourings are revealed after firing to a range of 680°C to 800°C.

John Wheeldon started printing with lustres after some years of making spongeware. The base glaze was overprinted with bold sponge shapes in contrasting brightly coloured glazes. When

Detail of printed lustre bowl, John Wheeldon (UK), 1993. Black firing clay with transparent glaze.

he started working with lustres, he brought the technique of the sponge print to bear on very different materials with correspondingly different results. Using an almost black firing clay body, with a transparent glaze in those areas to be lustred, he has developed a distinctive way of working with small stamps, often geometrical in design. Shapes are simple, but by repeating them, turning them 90° or 180°, he produces complex patterns with positive and negative printed shapes.

Prepared lustre is rolled out onto a ceramic tile. He uses ready-mixed lustres designed for painting, but by leaving them in the open for some time, the oils and solvents evaporate, leaving a stickier, 'fatter' lustre, suitable for printing with. (It is possible to buy lustres in a ready condition, for screen printing, rubber stamping etc.) If they become too 'dry', they can be 'let down' with lustre essence to the required consistency. He mixes different lustres to create different colours and qualities. In rolling out a colour ready for stamping, he may use two or three lustres rolled together, but not so well mixed that they merge completely. The effect is a tiny stamped shape, within it differing hues and iridescences. The effect of repeating this is to create subtle lines across or within forms, contributing as part of an overall effect.

The stamps are made from a dense, fine, foam rubber about 7 mm (4 in.) thick, 'from a Citroen engine gasket' is how he describes its origin. They are made by cutting the rubber with a scalpel, or for circular stamps, with a heated metal pipe (copper microbore, car aerials). Because the rubber sponge is porous, it holds a reservoir of lustre, allowing a number of repeated prints to be made from any one stamp before needing to re-ink.

Using sometimes similar techniques, Anne James also prints with lustres, but with quite different results. She works onto a white porcelain body, stained with sprayed coloured slips. Her forms, like John's, are thrown, but she burnishes to produce a smooth, softly shining surface. Her simple stamps are made from uphol-

Left: John Wheeldon's workbench with stamps, lustres, and 'inked up' tile ready for printing from.

Below: John Wheeldon decorating inside of bowl with lustre using stamp. Areas are delineated using masking tape.

stery foam, and the more complex ones from a denser foam. They are made by using conventional cutting methods and a soldering iron. (This gives off noxious fumes and should be done with good exhaust ventilation, or outdoors.) She also prints with crumpled plastic bags, cling film, polystyrene packaging, embossed wallpapers and fine, firm card for very fine lines.

The working method is to lay out a bold pattern with a strong painted lustre. Other lustres are subsequently printed over this to break it up, or make it much less obvious. A mother of pearl lustre over copper, for example, will actually partly mask it in one light, but in another will not be apparent. Several layers may be printed, with a firing taking place after each one, so that pieces may be fired six or seven times.

The firing itself is not conventional for after each lustre firing some smoking takes place. Here the pots are removed hot from the kiln and either plunged into sawdust, or have sawdust sprinkled over selected areas. Additional printing can

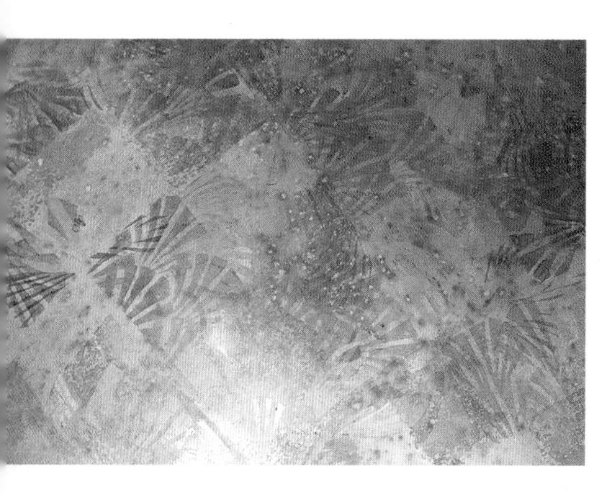

Above: Detail, bowl surface, Anne James (UK), 1993. Printed lustre and smoking.

Below: *Umbrella plate* Richard Slee (UK), diameter 30 cm. Lino print with rubber stamp border (underglazes) 1080°C.

also take place at the firing stage, using slip as an ink. This acts as a resist to the smoke, and after cooling, it falls away or is gently scraped off. The lustres are further enriched by these reducing and carbonising effects, which add to the subtlety and complexity of the finished surface.

Relief printing methods on ceramic surfaces are perhaps the most simple and direct, requiring relatively low tech materials and equipment. Their often inexact 'prints' are suited to ceramic processes which through fire often change, blur and soften printed images or designs.

Chapter Three
Intaglio printing

To a printmaker, 'intaglio' (from the Italian 'below the surface') means copperplate engraving, etching, aquatint and dry point. At first glance, it might seem that these disciplines have little to do with the studio potter or ceramic artist, but considering the rich tradition that the engravers of Stoke have left us, this is a pity. In addition, intaglio does not just mean printing from metal plates onto damp paper, but includes direct prints onto clay from metal, flexographic and plaster of Paris plates.

Prints from metal and similar plates

Copperplate engraving

Copperplate engraving, used by the pottery industry since the middle of the 18th century, has been refined to a fine art, with an elaborate vocabulary of symbols and textures, patterns and images. That most are stuck in a cultural and historical time warp, and the discipline of engraving very alien to the ceramist, means that serious consideration of designs, techniques and creative possibilities have not been explored. Engraving demands time, specialist tools and equipment, and can produce images and designs that are rather dead and mechanical, but this does not have to be the case.

A copperplate engraving is made first by planning the design using carbon paper, or drawing directly onto a metal plate with a wax crayon. This feint image is then made more permanent by going over the lines with a hard steel point. Next the lines are engraved with a variety of steel 'gravers' or 'burins'. The engravers with a lozenge section, cut V-shaped grooves in the metal plate. These grooves hold the ink when the plate is printed from. Other tools include punches, scrapers; burnishers, multiple gravers and gouges. Engraving is not something that can be easily picked up with a few minutes doodling, and a more specialist print book will better describe the process than there is room for here.

In Spode, a small group of skilled workers still make copperplate engravings, others printing onto pottery tissue and transferring to domestic ware. Most industrial engravings are now used with the Murray Curvex system.[1]

Dry point and etching

In contrast, dry point and etching are much more like drawing. There is no particular discipline of stamping or cutting to learn, no need for particular specialist tools. The making of lines or marks is basically by drawing, as you would with a pencil or pen on paper.

A dry point is made by scratching a design into a metal plate or other substance with a sharp-pointed instrument. This raises a furrow, rather like a plough

[1] See Chapter 1 *Historical and Industrial Background.*

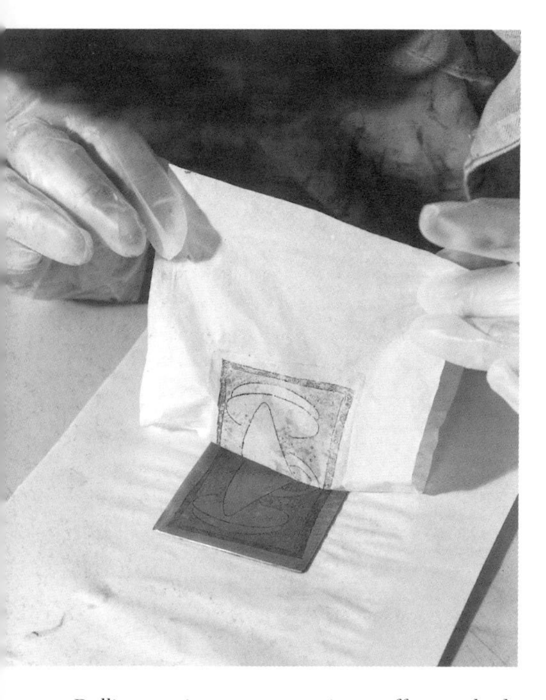

Pulling a print on pottery tissue off an etched plate. *Photograph by Andrew Morris.*

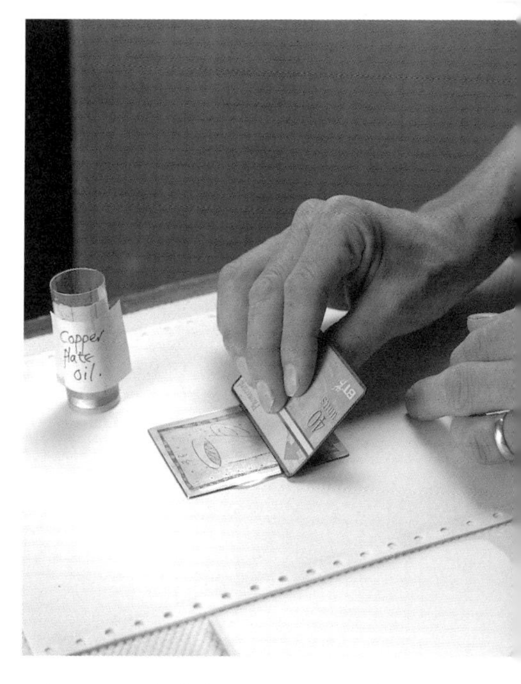

Inking up etched plate with copper plate oil. *Photograph by Andrew Morris.*

does in a field. Ink is spread over the surface of the plate, and cleaned off using a stiff canvas called scrim. The 'burr' holds the ink, not the indentation made by the scribe which is shallow and holds little pigment. The plate is put through an etching press in contact with damp paper, and the image transferred from the burrs to the damp paper. Because the burr is raised and the metal soft, dry point prints are generally small in editions, as the burr wears down with pressure from the printing press and the inking up process. Dry points tend to produce 'softer' prints than etchings or engravings. It sounds the most simple of techniques, and in some ways it is, but it does require a degree of skill and experience to get smooth flowing lines, and a little too much pressure can lead a line way off its intended direction.

To make an etching is to bite lines and textures into metal using a variety of mordants. First, a wax ground is rolled evenly onto a heated, polished plate of steel, copper or zinc. Alternatively liquid grounds in a solvent base can be applied with a brush. When the plate has been coated, the image is drawn onto the plate through the ground using a sharp implement. A scribe, needle or 6" nail will all make contrasting lines in an etched plate. Very little physical pressure is needed: the scribe glides over the plate even more freely than a pencil on paper. Plates are etched in mild acids or mordants, depending on the type of metal used and the qualities of line required.

Tonal areas can be created by using 'aquatints'. Traditionally, aquatints were created by dusting the metal plate surface with 'rosin' dust; the plate would be heated, melting the rosin which would form acid-resistant globules all over the plate

Placing gelatine pad on 'inked up' plate.
Photograph by Andrew Morris.

After placing the pad onto the ceramic surface, and pressing it into place, it is removed.
Photograph by Andrew Morris.

The image is revealed by dusting with ceramic pigment.
Photograph by Andrew Morris.

surface. An easy alternative is to use a car enamel spray paint to finely coat the plate surface. Whichever method used, by masking out areas and progressively biting the plate, a range of tones can be achieved.

Etching should be learned at a suitable print workshop. Traditionally the acids involved in biting plates, and some of the resists, are not 'friendly' materials and due care and attention should be exercised in their use and storage. More environmentally friendly materials and processes have made their way into print studios in recent years, but it is still wise undertake a specialist course in their use.

As with the dry point and engraving, prints are made by inking the plate, removing the excess, and printing in an etching press with damp paper. For ceramic prints, pottery tissue is used as the paper; oxides or underglaze colours thoroughly mixed with copperplate oil using a palette knife or glass muller, as the inks. Pottery tissue is applied to the ceramic surface after printing, and rubbed down, transferring the image. Although dry point and etched plates have been seldom used in the pottery industry, it is possible to obtain fire-able prints from both.

The dusting-on process, where printing takes place using oil instead of ink, and where the carrier is gelatine instead of tissue, is relatively simple from etched, engraved or dry point plates. Gelatine pads can be made from gelatine purchased in the supermarket or chemist. Using a stronger mix than normally recommended for food purposes, pour a gelatine mix onto a tile or glass plate enclosed by four card walls stuck to the surface with masking tape. When cool, the gelatine forms a pad which is removable with care. The pad may be susceptible to easy breakage at first, so some time should elapse before attempting to free the gelatine.

The pad will remain usable for some time, but eventually it will become brittle and hard. When working with an etched, engraved or dry point plate, copperplate oil is used as the 'ink'. The gelatine pad is placed onto the plate, and pressed firmly into place. On lifting the pad from plate, it is placed face down onto the glazed ceramic surface with slight pressure. The gelatine pad is then removed. A latent image now exists on the glazed surface which may be revealed by dusting on a ceramic colour.[2]

These methods, although tried and tested in the pottery industry do have limitations: The amount of colour needed for a print to work on a ceramic surface is greater than a paper one, and imperfect working procedures can make it a hit and miss affair for the individual, but they do work.

Flexography

Although Flexography[3] is primarily a relief printing process, it can also be inked for intaglio printing. As plates are photographically produced (intaglio use requires a film positive), the process lends itself to the production of digitally produced or altered images. Made from a light sensitive plastic photopolymer in differing degrees of hardness,on a steel or plastic sheet, it is probably unwise to ink these up on a hot plate, as the plastic plate surface is liable to melt.[4]

---

[2] Silicone pads made by extruding sealing silicone onto glass sheets might be usable instead of instead of gelatine.

[3] See Chapter 3 *Relief Printing*.

[4] See Chapter 5 *Computers, Screen Printing and Decals* for further details of film positives.

Test Tile, Andrew Atkinson (UK), University West of England. Four colour underglaze tissue print from intaglio flexo-plates.

Direct intaglio prints onto paper clay

Printing directly from a metal plate onto a clay surface (not possible according to some sources) is an exciting and rewarding process. It is difficult to create paper-thin sheets of clay, and usually these split on printing or stick to the plate, so for this approach to work effectively, printing must be done on 'Paper Clay'. Paper clay is a compound of paper pulp and clay, which has incredible green and dry strength. Upon firing the paper pulp burns away with no discernible effect on the body or surface.

Making paper pulp is relatively easy. Those papers made of cotton rag, or highly absorbent ones like blotting paper appear to work best, but cheap ones like photocopy paper will also work. Avoid highly glossed paper which takes a very long time to break down. Firstly, paper is shredded or ripped into small pieces and soaked in warm water to break it down. This should not be left indefinitely, as the mixture will soon take on a pungent odour. After mixing in a blender, or using a glaze mixer, the resulting pulp is squeezed to remove as much water as possible. At this stage pulp can be bagged and frozen if not all of it is required at once. To make paper clay, the pulp is mixed one part pulp to two parts slip (I use porcelain) in a blender or liquidiser. The resulting slop is spread evenly over a plaster slab to dry.

Mixed paper clay should be turned into clay sheets as soon as practical. The mixture does not store well and turns distinctly aromatic after a short time. Dry paper clay (scraps etc.) can be reclaimed as per normal clay reclaim, but again, avoid too long a period of paper clay being a liquid state.

Spread paper porcelain sheet, showing guides used to ensure even thickness of paper clay. Also a paper-thin dry sheet is held showing its relative strength. *Photograph by Andrew Morris.*

Large, paper-thin sheets of paper clay can be made without fear of warping. The product has incredible green and dry strength, and pieces can be stuck to each other dry, or dry to wet (just using a slurry of paper clay). Bisque- and glaze-fired strength is the same as for normal clay (so paper-thin sheets of fired paper porcelain are still extremely fragile). The material has many creative possibilities, and is currently being used by ceramic sculptors with great enthusiasm.

Paper clay makes possible the printing of etchings, dry points and engravings directly onto a ceramic surface. Ink is made up by mixing copperplate oil and underglaze colour (or oxide) using a glass muller or palette knife. The ink should not be too thick so that it is sticky and lumpy, but not too liquid so that it runs off the palette knife. Plates are inked up using an old credit, telephone or bank card. Traditionally, plates are inked on a hot plate, but it is possible with the right ink consistency to do away with this. Most of the 'wiping' can be done with the edge of the card, with a final cleaning with a small sheet of newsprint to polish

'Pots No. 40' by Paul Scott, 12 cm x 17 cm. Etching on paper porcelain with underglaze colours and transparent glaze, 1200°C. *Photograph by Andrew Morris.*

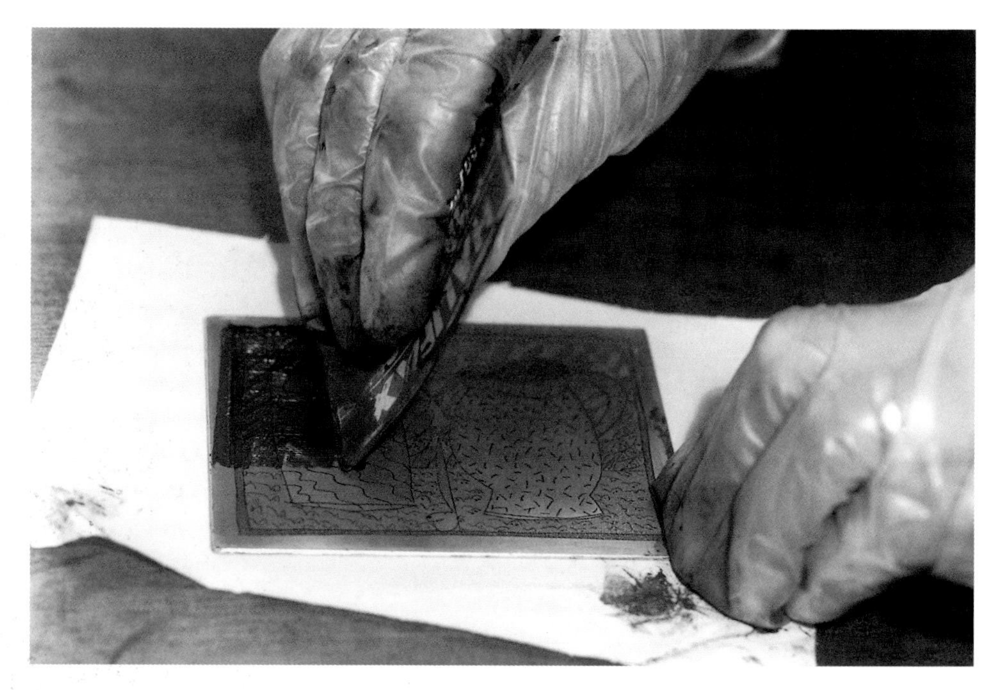

Inking up etched zinc plate with ceramic ink. *Photograph by Andrew Morris.*

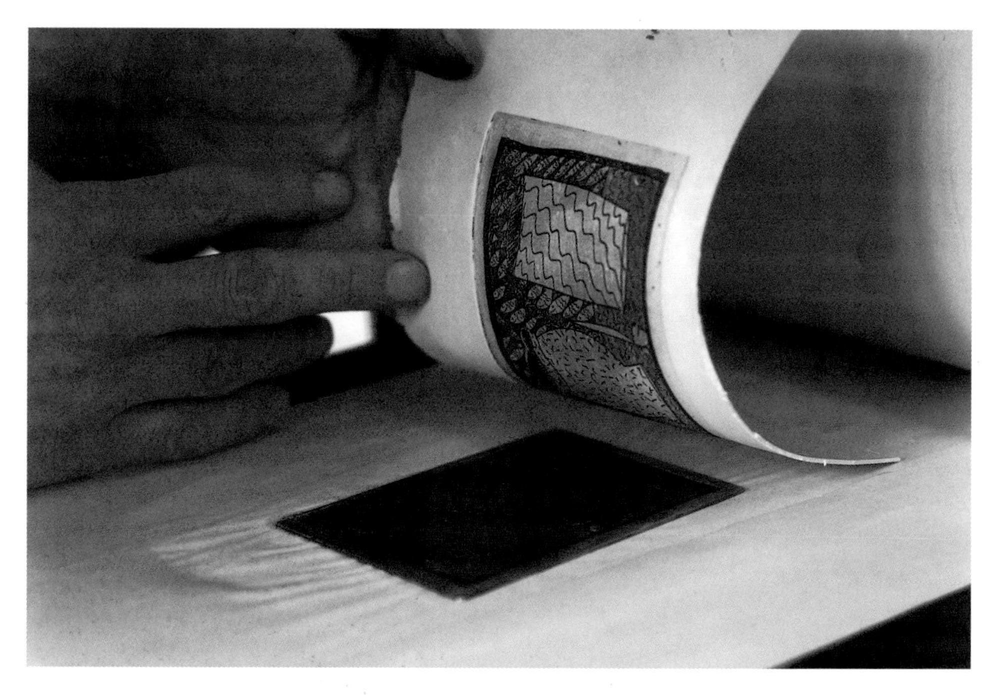

Lifting paper clay from plate after passing through press to reveal print. *Photograph by Andrew Morris.*

the plate. The object of the exercise is to have the maximum amount of ink in the grooves of the plate, with a minimum on the plate surface. When the plate is inked, it is placed in an etching press, and a leatherhard sheet of paper clay placed on top. Adjusting the pressure of the press is important to achieve a good print. After passing through the etching press, the paper clay can be lifted to reveal the print. Depending on the underglazes and oxides used, a wide range of drawn and ceramic qualities can be achieved.

Prints from plaster

The ceramic printer has a material so versatile that the paper printer might well be envious of it: Plaster of Paris. Plaster can be easily drawn into with a sharp scribe. Not perhaps as easily as drawing on an etching, but then there is no acid to handle. Easier than engraving

Lotte Glob (UK) Copper plate etching printed onto leatherhard paperclay, with ceramic pigment. Fired onto a thin crank tile with clay and rocks from the local river bank, which start melting and fusing the two together. Globb has recently been using her paperclay prints in ceramic books.

in metal, or making a dry point, plaster is best worked soon after it has set.

For the finest prints, high density plaster is best used. Although it is the least porous, and so takes longer to release wet clay or slip, it is made of very fine particles of plaster, and will give finest detail. Slabs should be cast up to 3 or 4 cm (approx. 1 ½ in.) thick depending on the size of print. Designs are drawn or carved into the plaster surface. Prints are taken by beating thin sheets of clay into the plaster surface.

Mo Jupp used this technique to produce prints for the Miniature Print

State of Play #5, Mo Jupp (UK). Backlit porcelain intaglio print from plaster, 1260°C.

Left: Porcelain intaglio print next to plaster block with scribed image, by Paul Scott. *Photograph by Andrew Morris.*

Below: *Banana Flower Bowl* Claudia Clare (UK). Intaglio plaster print slip cast from drop flower mould.

Joanna Veevers (UK), detail intaglio ceramic monoprint with stained slips.

Collection at Fine Print Research Department at the University of the West of England.

In a different way, Joanna Veevers produces intaglio monoprints from carved plaster blocks. Her delicate, graphic qualities are achieved by drawing fine, incised lines in plaster, which are subsequently 'inked up' with black slip. The block is wiped clean leaving black in the incised areas. Further detail and colours are built up with painted washes and layers of slips. When complete, the image is held together with a layer of backing slip. When in a suitable condition the print is peeled from the plaster revealing the fine drawn line, colour and detail. Her work has a fine, graphic quality not often associated with ceramic surfaces.

Chapter Four
Monoprinting

A monoprint or monotype is a 'drawing with printed features' or print in an edition of one. Monoprinting is said to have been first used by 17th century Italian, Giovanni Castiglione, and both Degas and Matisse made monoprints. Because a print of one is of little use in mass production it has been of no significance in the ceramics industry, but a number of monoprinting techniques are possible using ceramic materials

Monoprinting from plaster and fabric

Plaster of Paris is a material well used by the ceramist, and ideal for monoprinting from. Like Mo Jupp, but preferring the solitary print as opposed to the edition, Joanna Veevers' intaglio prints are actually monoprints because she only pulls one per plate.

Another use of plaster is made by Les Lawrence, who screen prints with slips onto a plaster plate before pulling the image off in the surface of paper thin poured slip slabs .[1]

Jerry Caplan uses 'Reward Velvetone' underglaze colours as printing inks. He paints thickly onto a damp, canvas covered board until satisfied with the image or design. Then a slab of well prepared white stoneware clay is placed on top, and gently rolled with a rolling pin. On removal, the painted design is revealed in reverse on the clay surface. A second slab laid upon the canvas will also take up the design, more faintly but differently. Sometimes the second 'monoprint' can be more effective than the first. The printed slabs can be used simply as clay pictures, but also by using moulds they can appear as plates or shallow bowls. Annie Turner also uses fabric to monoprint with, using coloured slips, painted onto cotton and other fabrics. For her, the printing is integral to the forming and building process.

New Visions Vessel, Les Lawrence (USA). Porcelain – photo/silkscreen transfer, height 35.5 cm (14 in.) x 7.5 cm (3 in.) dia., 1991.

[1] See Chapter 5.

Terence Bond, *Family Ware (Detail, 1 of 4 parts)*, 1999, 27cm (11 in.) dia. Monoprinted enamel on bone china.

Hanne Heuch (Norway), commission for RIT 2000. Installed in the radiotherapy unit Trondheim Hospital, Norway. Porcelain monoprinted slips from plaster slabs. Heuch has been monoprinting from plaster and fabrics since the early 1980s, often for large-scale architectural commissions.

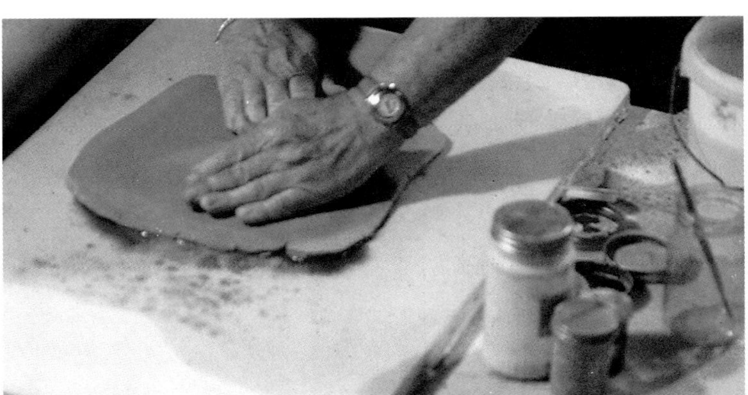

Above: Jerry Caplan (USA) Placing blank sheet of clay onto painted master.

Left: Rubbing back of clay slab placed over pre-painted canvas covered board.

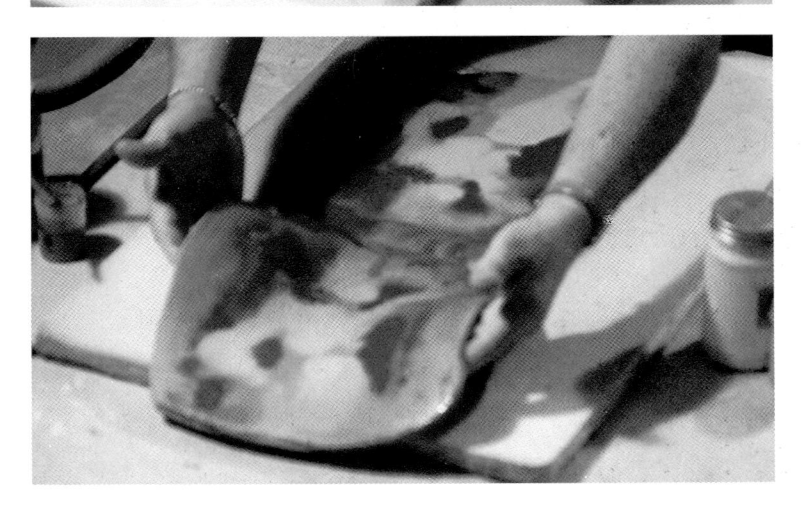

Left: Removing slab to reveal mono print.

Finished pieces awaiting firing. Jerry often uses this monoprinting technique in conjunction with raku.

Detail of large plate, by Annie Turner(UK). Painted then hump-moulded, then over-monoprinted from cloth with slip, earthenware.

MMI #2 Mitch Lyons (USA) 40cm x 66cm (16"x 26"). In a curious reversal of ceramic print practice Lyons uses stained slips to print with, but the finished artworks are on paper. Using a large china clay slab (now 21 years old) as the plate, Lyons works into its surface with slips stained with various ceramic and paint pigments, pulling his clay monoprints off on sheets of paper which are dried and displayed as 'conventional' paper prints.

The Message Sara Robertson (UK). 'I paint a base colour of slip onto paper building up layers and then waiting until it is soft but not wet. Using a variety of tools and textures I draw into and mark the slip before painting over the base colour(s) with the colour(s) which will show through the lines and marks. When the slip still has a sheen on the surface but is not wet I apply it to a pre-rolled slab of clay. The image prints in reverse. Using this method it is possible to over print building up texture with the layers of slip. Often I paint, scratch and draw into the slip after printing. This can create a rich surface.'

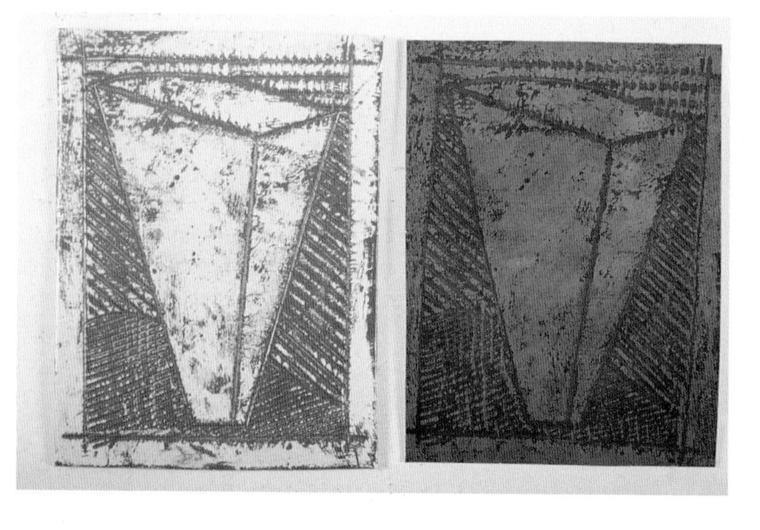

Clay Monoprints, Orange Mono, Jackie Brown (USA). Clay Monoprints each 33cm x 43cm (13"x 17"). Monoprinted from slip coated cotton fabric.

Bottles Fiona Thompson (UK), tallest 29cm (11.5"), 2000. Hand-built paperclay porcelain monoprinted slips (from newspaper), further images drawn in with further slip added then transferred onto clay surface. Additional lines monoprinted on top. Matt and transparent glazes applied. *Photographer, David Gilroy.*

Kiln prints

The Swedish ceramist Herman Fogelin has referred to the kiln as a sort of washing machine.[2] Caplan and others use it as a printing press, imparting colour and other unique qualities. The kiln, in generating heat, can facilitate either the transfer of colour from one body to another, or causes a print by reducing clay and carbonising matter.

Caplan coats bisque fired ceramic surfaces with slip, and then draws through the drying surface. During raku firing the fire draws oxygen from the surface of the ceramic through the inscribed lines of coated slip, thus chemically reducing the clay. Later removal of the unfixed slip resist reveals the drawn image in dark, bleeding lines on the lighter (oxidised) clay surface.

Dick Lehman's carbon film transfers on sagger fired porcelain use the same basic chemical principles as Caplans, but in this case the materials, process and intention are different. 'Typically a sagger is partially filled with 5inches (13cm) of sawdust and pressed down to create a 'nest'. The pot is removed and fresh vegetation is positioned in the nest. Then the pot is put back in place-on its side atop the vegetation. Next more vegetation is placed onto the exposed top side of the pot. It is then covered by an additional 5 inches of sawdust. The saggar is then closed with a lid.' In the firing a *reducing* (or *anaerobic* – no oxygen) atmosphere is created inside the saggar, and the vegetation turns into activated charcoal, in the process releasing a film of carbon, which the bisqued porcelain absorbs, capturing the image released by the vegetation.

---

[2] Herman Fogelin quoted in *Painted Clay, graphic arts and the ceramic surface*, Paul Scott (A&C Black 2000) p. 118.

Although Lehman observes that the success rate is less than 20%, the wonder and surprise of the results that do work propel him to continue exploring the process.

Janet Williams has also worked with the firing process and organic matter to produce a variety of 'printed' images in refractory cement, latex and clay. She is concerned about how firing alters organic matter, carbonising it, with the residues forming the focus of some pieces.

Many ceramists will be aware of the ability of some ceramic pigments, usually in oxide form, to volatilise in the kiln and transfer colour from one piece of work to another. Usually this is accidental, and not always desirable, but Peter Beard has made particular use of this phenomenon to make kiln monoprints (see photo-

Left: Jerry Caplan (USA) drawing through slip resist on bisque piece, *Roman Fresco.*

Below: *Roman Fresco* Jerry Caplan (USA). After firing, slip stencil removed.

Trio of Saggar Fired Vessels, Dick Lehman (USA), grolleg porcelain, tallest piece height 23cm (9"). 'It was an odd and curious experience to see on my desk each day successful veggie saggar pieces – pots with delightfully delicate imagery, pots with such explicit detail that I could see the veins and tears and worm holes in many of the leaves, pots that were the ceramic counterparts to the contact prints I routinely made from my 4x5 negatives in the darkroom.'[3]
Photograph Dick Lehman.

[3] Dick Lehman in *Ceramics Monthly* March 2000, p 35.

Detail, *Prarie Flora: Leaves* Janet Williams (UK/USA), 1996,Ceramic, impressed flora, carbonised residues in wooden frame with leather ties.

graphs on p. 77). Using two porcelain slabs the same shape and size, he incises one with a design when leatherhard. This is then covered with a thick 50/50 mixture of copper oxide and China clay, ensuring that it is forced into all the grooves and incised lines. A subsequent scraping with a steel kidney reveals the design. The slabs are laid face to face and biscuit fired (1000°C). After firing, the blank tile will have a faint image in reverse of the original, which will appear dark and strong. After applying a glaze which responds well to copper, and refiring, the results are two similar objects each patterned, but with different qualities. Other oxides may well work in this process, and the method of application, firing and glazing will each affect the finished quality of work.

Powder Monoprints

Back to more conventional print processes, a remarkable quality of drawn line is achievable using powdered colour. A sheet of glass is painted with oxide or underglaze colour simply mixed with water. When the colour has dried, a sheet of pottery tissue is taped on all four sides onto the glass. This prevents the paper moving thus keeping it taut and flat. Dry drawing (i.e. with pencil, biro, chalks) onto the paper works best, but the surface is very sensitive and will also pick up from drawings done with a cotton bud for example. It will also pick up unintentional marks from finger or hand pressure, so it is important not to touch the paper surface more than absolutely necessary. When the drawing is complete, it can be cut out of the masking tape holding it in place with a scalpel, and placed face down onto a damp clay surface. After rubbing the tissue gently, but firmly all over, it is removed. The drawn image will

Above: Tile showing inking up process using copper oxide and china clay. The top of the tile has been filled and scraped back, the middle of the tile shows the copper mixture filling the lines, the bottom of the tile is uninked.

Above: Tiles after biscuit firing. Tiles on the left are the monoprinted images.

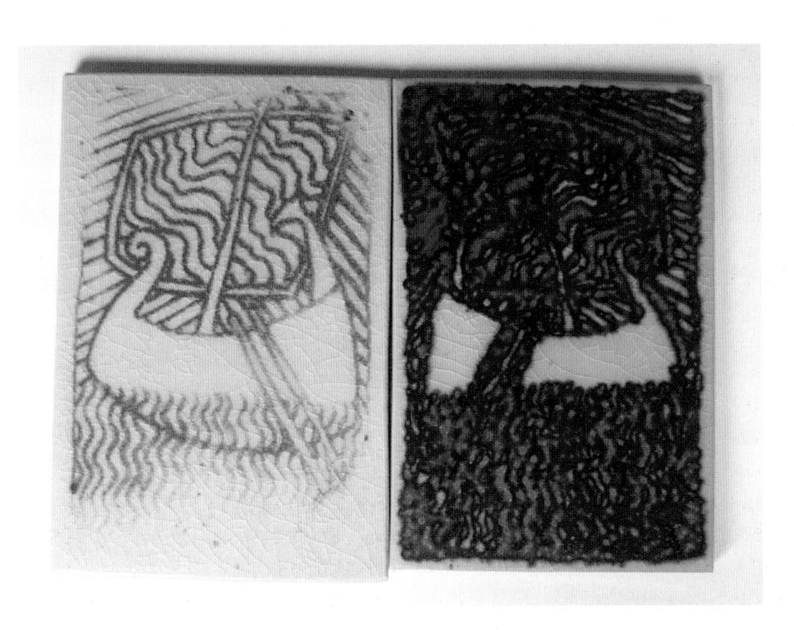

Left: Kiln monoprinted tiles, by Peter Beard (UK). Celadon glaze, reduction fired.

have transferred onto the clay surface. Different drawing implements will produce different drawn qualities. A pencil line will be quite different in character from a biro line or a soft smudge from a finger. It is possible to monoprint several layers of colour.

The process is simple, and does not rely on exacting conditions, but a number of factors should be borne in mind. Oxide and colour should not be mixed with an underglaze medium. This binds the colour too strongly to the glass or tile, and an effective transfer is very difficult. The clay should not be too wet or too dry. Wet clay will distort and disintegrate the tissue, and dry clay may not take the print as well. There is a tolerance of clay conditions though and experience will tell when the body is just right. Used glass plates and tiles should be wet cleaned after use, or stored with a covering of cling film to reduce dust generation. A wet brush painted over the surface of the tile or glass will restore the monoprint plate for further use. Karen Densham

Women's Institute Ware (Breast Plate) Karen Densham (UK), 27cm (10.5") dia. Ready made plate with powder monoprint onto fired glazed surface.

uses monoprinting in this way on her plates, combining the technique, with painting and stamping.

Richard Slee has used monoprinting by rolling or painting on glass using fat oil and pure turpentine as the carrier for the ceramic pigment. When the image is complete, it is transferred to biscuit using pottery tissue. The back of the print is rubbed down with fingers, soft soap acting as the lubricant to avoid tearing the tissue.

Remarkably, it is possible to monoprint from photocopies, and in some cases the photocopy actually fires a sepia colour. In order for the process to be worked, one must have internal access to a photocopying machine.

Photocopiers (and laser printers) work on the natural phenomenon of static electricity. Negatively charged dusting powder (toner) is dusted over a photosensitive surface which has been exposed to a projected image of the document or image to be photocopied. It clings to those areas of the image which are positively charged – the darkest areas. A sheet of paper is passed across the photosensitive surface (usually a drum), and a positive electrical charge below the paper surface attracts the toner from the drum to the paper. At this stage the image on the paper is in fine dust; it is fixed by passing it through heated rollers which softens the toner and fuses it to the paper fibres.

Peeling off tissue print showing transfer of drawn image.

If the photocopying machine is turned off just before the paper passes through the heated rollers, and the paper removed, the image can in turn be trans-

Test tissue monoprints by Ane-Katrine von Bülow (Denmark) created by overprinting and layering.

Polished porcelain form with monoprinted and painted detail. Ivana Roberts (UK/Croatia). Roberts uses paper monoprints in combination with rubber stamping, decals and painting to produce the images on her handbuilt and polished porcelain forms.

ferred to a clay surface by simply laying the paper face down onto a prepared slab of clay, and rolling gently. Upon removal, the image appears on the clay surface. On firing, some photocopy toner will sustain a ceramic colour,[4] generally sepia. Other toners disappear completely, but if the photocopy has been sufficiently well rolled onto the soft clay surface, the image can be restored after biscuit firing by washing the ceramic surface with an underglaze colour, and wiping with a sponge. The photocopied image magically appears from a seemingly blank tile.

The iron oxide in some photocopy and laser printer toners not only allows the production of monoprints from unfused prints or copies but can be used in the production of decals.[5]

---

[4] Copiers and laser printers vary in their toner constituents. Some Apple, Canon and Hewlett Packard branded machines do contain iron oxide. On the other hand, Minolta copiers, Epson laser printers and NEC machines do not appear to. A guide to toner content can be gleaned from specific toner Health and Safety data sheets (or MSDS reports in USA).

Trowel and Crescent monoprinted plate, by Richard Slee, 30 cm diameter, 1080°C.

[5] Even fused photocopies can be transferred to fired tiles. Place photocopy face down on glazed tile and iron the back firmly. The heat of the iron will melt the toner and stick it to the glazed tile surface. Paper can later be removed by soaking in water. On firing any remaining paper is burnt away and the print adheres to the glaze.

Un-covercoated photocopies on decal paper can be transferred to clay by back sponging with decals whilst face down on clay or bisque ware that has been 'pre-stickied' with spray mount. Firm pressure is needed and most success is achieved by using a spoon or similar tool to burnish the back of the paper.

Where Have all the Flowers Gone, Warren Palmer (Australia). Ceramic wall tile, fired photocopy with mixed media. Palmer transfers his drawings and images to a ceramic surface via powder photocopy, later enhancing the printed image with ceramic colours and lustres.

Above: *Cumbrian Blue(s) Tree*, Paul Scott, powder photocopy monoprint on porcelain 1200°C, 8cm x15cm ((3"x 6").

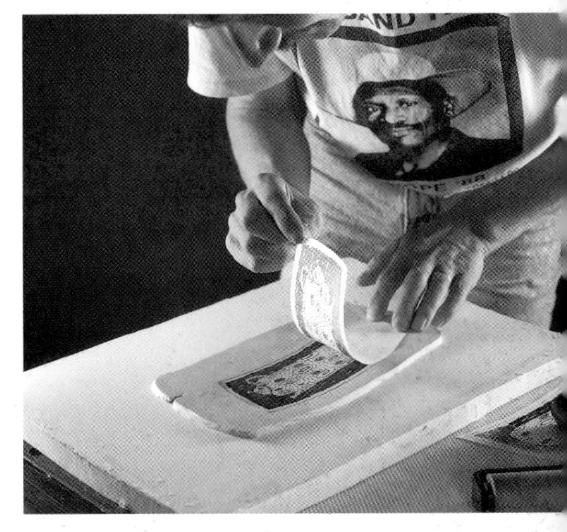

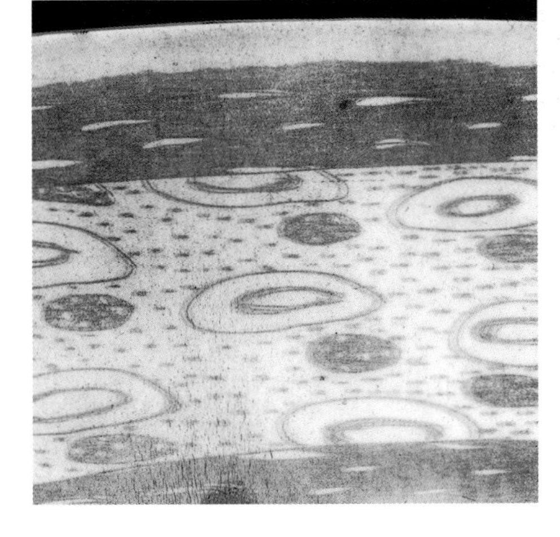

Right, top: Rolling photocopy onto slab.

Centre: Removing photocopy revealing print.

Bottom: Detail of fired surface of porcelain slab, where photocopy had disappeared after biscuit firing. It was returned by painting biscuited slab with black underglaze, then wiping clean with a sponge. *Photographs by Andrew Morris.*

Chapter Five
Photocopiers, laser printers, computers, screen printing and decals

Decals

Although decals are a relatively simply and flexible way of putting print onto a ceramic surface, they can also facilitate the application of complex, precise, and multilayered prints. There are two basic methodologies in use:

Over covercoating method: traditionally involves printing an image or design in oil-based ceramic inks onto a specially gummed *decal* paper.[1] Lithography, lino-cuts and screen printing are the most

[1] Trade names include Tullis Russell's *Twincal*,(UK/USA) *and Simplex Sta-flat* paper made by Nazdar (USA).

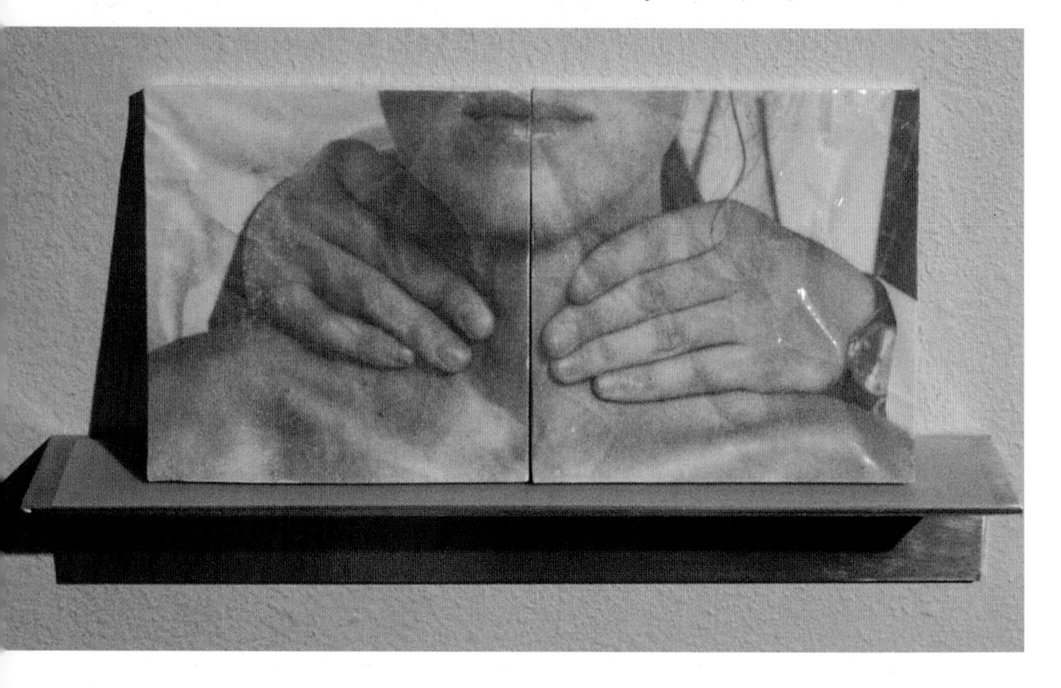

The Nurse, Katherine L. Ross (USA), porcelain, latex, aluminium, with fired laser decal print. Ross is only interested in one or two quality reproductions of her images, and although screen printing offers quality, it is also time consuming and resource hungry if a print of one is all that is required. As a result she uses laser decals, accepting and using the restricted sepia colour it fires.

Once Upon a time in the West, Robert Dawson (UK). Bone china plate with laser decal print, 2000. Dawson's work cleverly plays with our perceptions of objects, and their uses. He frequently alludes to the duality of the plate, as functional object, and on the wall as art.

common ways of making decals, but it is also possible to put conventional decal paper through laser copiers and photocopiers which have iron rich toners[2] to make ceramic prints.

The paper with its printed image (oil-based ceramic inks, or fused iron oxide toner) is then coated with a liquid lacquer or varnish called *Covercoat*, which on drying becomes a thin plastic sheet including the printed image. Placed in warm water, the plastic sheet including the print, becomes moveable, and can be slid off the gummed paper and positioned on a ceramic surface (usually glazed or polished, or if bisque, sealed with gum arabic). On the glaze it can be slid around until correctly positioned, then firmed into place with a rubber kidney, or by

[2] See p. 81 for details.

New Vision Plate, Faux Lennox #4, Les Lawrence (USA). Laser decal, china paint (painted), on recycled Wedgwood bone china plate, 1999, 30cm (12") dia.

rubbing with a finger, ensuring that no pockets of air or water are trapped below the covercoat surface. The residues of the gum help the transfer to adhere. Upon firing the plastic burns away, leaving the printed image or design on or in the ceramic surface (always ensure good exhaust ventilation on firing).

Covercoat can be applied by painting or using a plastic credit type card to pull a thin layer of lacquer across small prints, however it is preferably applied by squeegeeing through a blank, relatively open silk screen. The lacquer should only be used in well-ventilated (preferably air extracted) areas. Prints also need good ventilation as the vapours are potentially explosive if contained, and screens need cleaning with solvent thinners.

Attempts to develop a more user-friendly lacquer material have all

Right: *The Scott Collection, Cumbrian Blue(s)*. Paul Scott (UK) 1999. Inglaze decal screen-print collage on bone china plate.

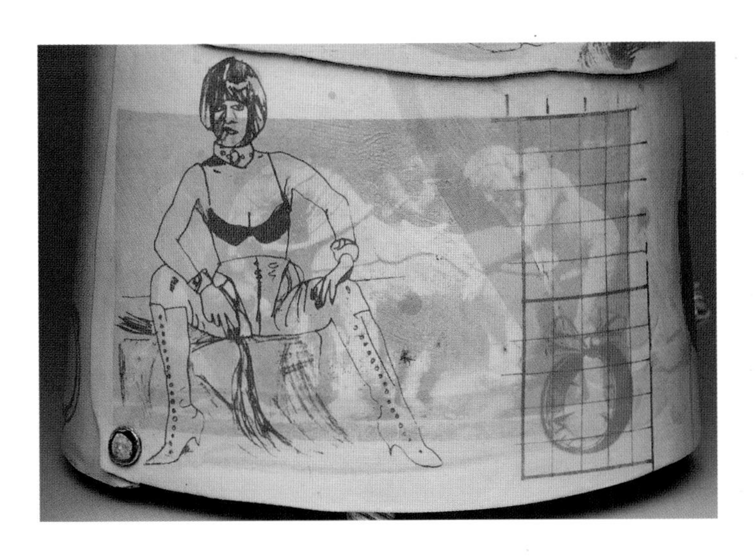

Left: *Decadence*(detail), (Private collection USA) Stephen Dixon (UK) 2000. Photocopy transfer print, and overglaze water-based decals. Sometimes deliberately high-fired to fade images, Dixon uses a combination of print processes and firing temperatures to develop complex and subtle surface qualities.

Left: Jean Oppengaffen (Belgium), porcelain boxes with screen-printed decal detailing.

Below: *Cylinder with blue steps,* Bodil Manz (Denmark). Slip cast porcelain form with screen transfer printed surfaces

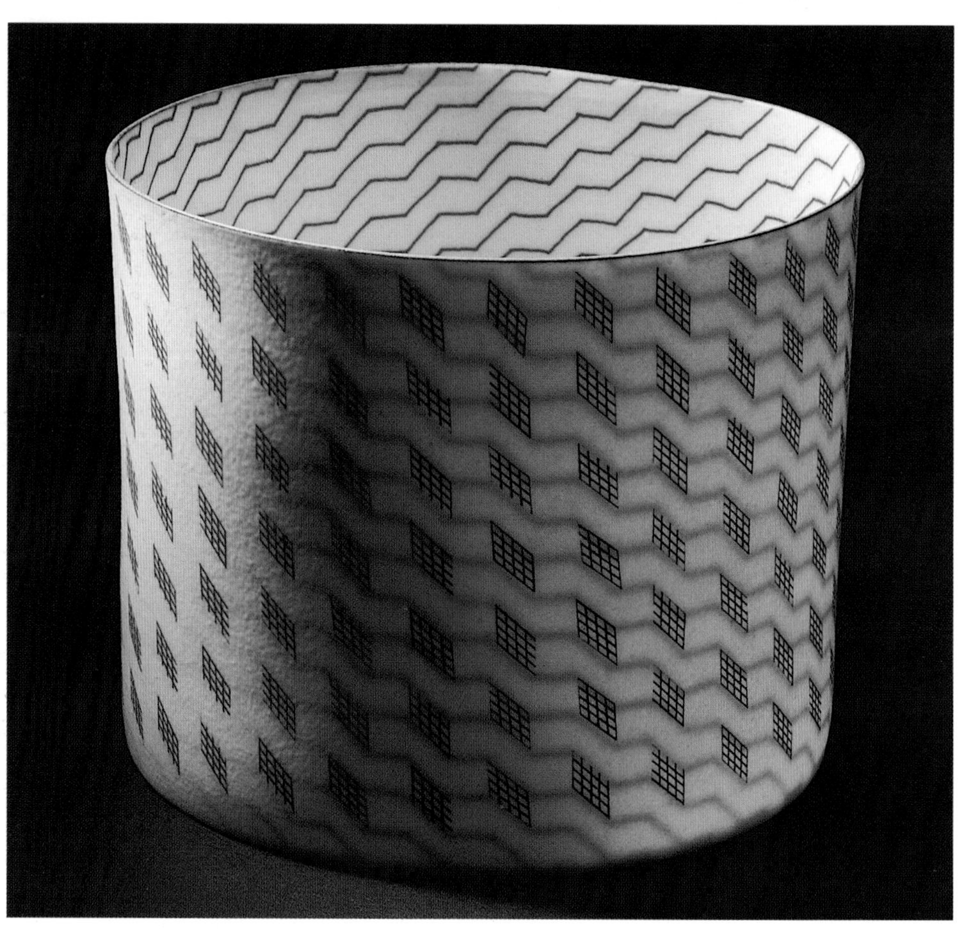

foundered. The material has been specially designed to work: to lift an image off gummed paper in water, to be flexible in application to glazed ceramic surfaces and most importantly to burn out cleanly without distorting the printed image on firing. Substitutes have been researched but to date nothing appears to work as well as covercoat. [3] However, the efforts to develop more user friendly technology have produced different systems using pre-coated paper in recent years....

[3] Alternatives used by artists have included spray lacquer from the hardware store, (Katherine Ross recommends up to six sprayed coats of Krylon clear acrylic paint) and spray on skin as used in the medical profession. Unlike covercoated decals which will store almost indefinitely, most require relatively quick application (within two or three days) before they become brittle. Like all non-conventional uses of materials, any exploration of these alternatives should be undertaken with extreme care, and in particular as with conventional covercoat, ensure good exhaust ventilation on firing.

Printing on pre-coated paper: Several types of pre-coated decal paper are available from a number of manufacturers. These products are very specifically produced for particular methodologies and technologies, and experimentation with them outside their intended usage should only be undertaken with extreme care.

UWET paper has been designed for use with a water-based screen system. Avoiding the need for printing covercoat, (and all the mess and fumes that go with

Walking the Dog Kevin Petrie (UK), overglaze water-based decal print on bone china plate 1998. Petrie uses the plate as a picture plane in the same way as the paper/printmaker uses paper, but he also makes reference to a tradition of *trompe l'oeil* ceramics.

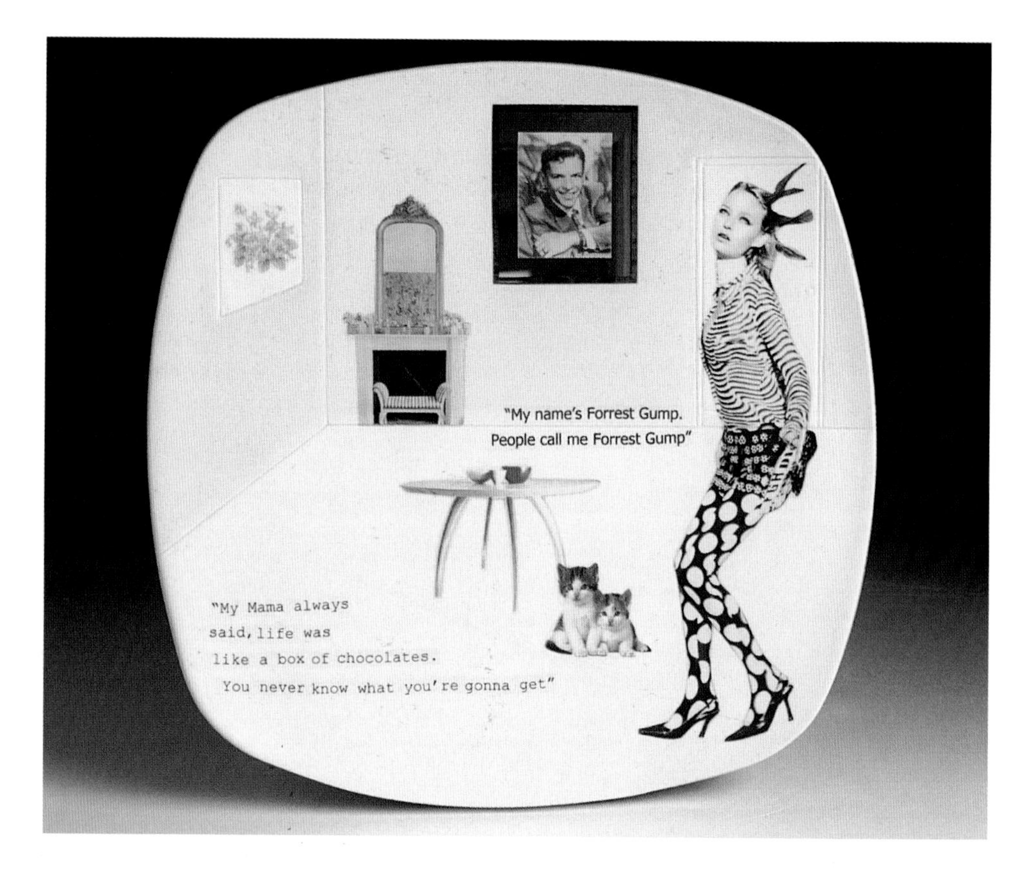

The Forest Gump Obsession, Helen Bromby (UK). Inscribed slipcast earthenware plate, vellum glaze, with lazertran decals refired to 180°C in conventional oven to seal.

it) images and designs are simply printed on top of the pre-coated paper in a specially formulated water-based ink system. When the ink is dry, decals are usable in the same way as those produced conventionally. On firing the covercoat burns away from underneath the printed image.

Lasertran paper is designed to be used with certain colour laser printers and photocopiers.[4] The resulting decals are applied to the glazed surface in conventional manner but then 'fired' in a conventional oven(180°C) to seal and stick the print to the glaze surface.[5]

Other non ceramic ink technologies including inkjet printing with sublimation inks are applicable to ceramic surfaces but like lazertran are not normally fireable. However, Greg Daly (Australia) has successfully used inkjet printing, making 'ceramic' inks by injecting a nearly empty inkjet cartridge with soluble metal salts (*e.g.* cobalt nitrate, or chlorides, sulphates). **Special care should be**

[4] See www.lazertran.com for technical details and product specification before using in any machines.

[5] Never use UWET or lazertran in mono laser copiers or mono photocopiers as the fusing temperatures in these machines is in excess of the melting point of the covercoat on these papers. Serious damage to expensive hardware could be the outcome of careless experimentation with these papers.

Binding site 2000, Jefford Horrigan (UK). Clay fabric and laser print. Horrigan simply uses conventional laser prints, sticking them to his clay and fabric sculptures with water. They are unfired. *Binding Site* is part of a series of box hedge-like structures that reflect the binding sites of cells. The viewer moves between the work as if walking through a formal garden.

made to observe relevant health and safety precautions with these materials as they can be hazardous. The salts are dissolved in the ink which was transferred from printed paper by simply rubbing face down on a damp clay surface, Other printing alternatives include direct printing with thermoplastic colours that are bright and durable, but require specialist equipment.

Screenprinting

However instant, colourful and flexible these laser decals are, they are not strongly fused to the glaze but an adjunct to the surface, that can be removed by scratching with a metal scraper. Covercoated laser and photocopy prints on the other hand can be fused to the glazed surface, if fired high enough, but

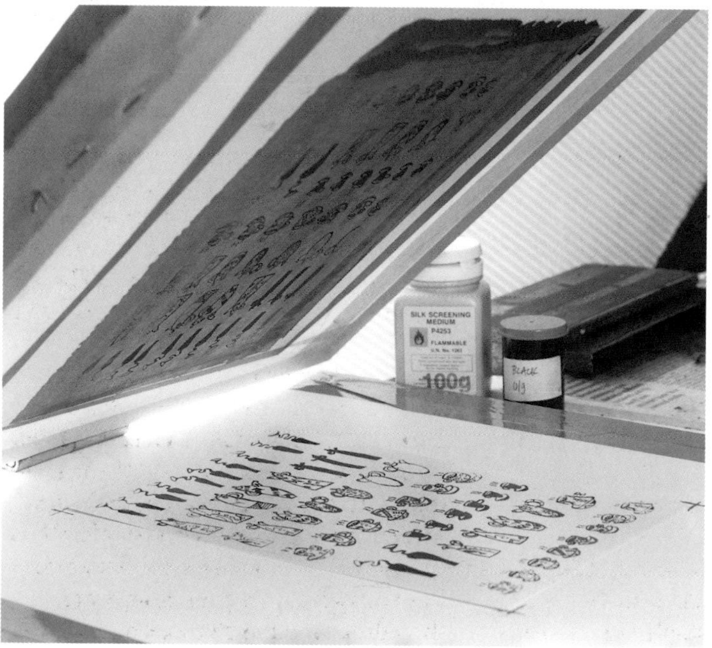

Above: Printing onto decal paper, pulling squeegee across silk screen. *Photograph Andrew Morris.*

Left: Lifting screen to reveal print on decal paper, in this case the paper is pre-covercoated and so only requires the ink to dry before use. Using normal decal paper, the print would require overprinting with covercoat. *Photograph Andrew Morris.*

After cutting out the required image, it is soaked in warm water, then the decal is slid off its backing paper onto glazed ceramic surface. *Photograph Andrew Morris.*

the resulting print is always monochromatic and sepia.

For permanent colour in either underglaze or overglaze, photocopiers and laser printers are limited without an engagement with other printing technology. Allied with the computer, screen-printing is the most flexible process for decal production.

Silk screens are frames of wood or metal, one side covered with a stretched fine mesh made from silk, polyester, nylon or steel. Designs are put on screens in substances that effectively block the mesh. These are transferred to paper, clay or other surfaces, when the silk screen is placed over the surface, and a 'squeegee' pulled across the inside of the screen forcing ink through the open mesh onto the receiving surface. Where the screen is blocked no ink appears, in open areas ink does.

This is the basic principle. In practical terms, screens for ceramic use can be made cheaply in the studio from basic materials such as (mitred) slate battening, nails, mesh and a staple gun. Screens should be larger than the images to be printed, allow at least an extra 10cm border around image size ensuring space for ink and squeegeed colour. If your wish is for precise, controlled, colour registered multiprints, or for fine quality detail then more expensive and professionally produced equipment is called for.

Types of mesh

Glancing at a print supplier's mesh listing can be extremely confusing, but for most common, studio, print and ceramic use, monofilament polyester is the best material to use. It is tough, dimensionally stable, and designed specifically for the job it does.

Mesh sizes are measured in threads per inch or centimetre. Strength for strength, the traditional screen material, is thicker per thread than synthetic materials. The effect of this is that each opening in the mesh in the silk is smaller than the monofilament polyester, so there is less open area for a given number of threads per inch: A silk screen of 195

Untitled Vicky Shaw (UK) 1994. Screen printed underglaze, glaze and decals porcelain, perspex.

mesh (195 threads per inch, or 77 threads per cm) will have an open area of 24 per cent, whereas an equivalent polyester screen has an open area of 32 per cent allowing a heavier deposit of pigment.

Screen printing can be as basic as using paper stencils or masking tape to block out relatively large or crude areas, or as sophisticated as having multi screens printing four-colour prints from computer-originated images.

Deciding on a mesh depends on the level of sophistication of the designs and printed quality required. Bold, large shapes with little detail will probably be best produced on a screen of 110 threads per inch (43 threads per cm), whereas for photostencils using half tones a finer screen of 230 threads per inch (90 per cm) should be more suitable. However, ceramic pigments are generally coarser than ordinary inks. Underglaze colours for example are routinely passed through an 80s mesh (80 threads per inch) before sale, although overglaze colours are ground finer to at least 200s mesh. Specialist books advise that mesh size should ideally be at least 2.5 to 3 times the size of pigment particles. So using underglazes through a 230s mesh screen doesn't work, the screen blocks and only a very thin layer of colour is deposited. If you need very fine detail, use overglazes, or have inks professionally mixed for you. If you do make your own, when ordering colours specify that they are for screen printing, and give the mesh size and ensure that colours are well mixed with media.

Putting designs on screens

The simplest screen print is from a blank screen. Screening ink through a blank screen will produce an area of colour. By masking areas with masking tape or paper, un-inked areas will appear.

Vicky Shaw uses blank silk screens, adapting them with masking tape and paper, to screen glazes and ceramic pigments to make her ceramic abstracted prints. Using water-based media, she exploits the screen's tendency to block with ceramic pigments and glazes, sometimes forcing colour through the screen with palette knives, sometimes drawing onto the printed surface before firing. Prints are made on thin sheets of porcelain, and are multifired. The qualities of the prints are exclusively dependent on the ceramic pigments, glazes and the firing process. They are qualities unobtainable by using paper and conventional inks, but they are undoubtedly 'prints' as opposed to pieces of pottery.

More elaborate prints can be created by using a blocking out medium, and painting this directly onto the screen. Unless the desired effect is the image degrading on printing, the blocking out medium should not be dissolvable by the print medium i.e. oil-based inks should mean a water-based blocking medium.

Neal French (UK)has used a basic set of patterned screens for many years. With a basic mix of 50/50 China clay and ball clay mixed to a yoghurt-like consistency and oxides for pigmentation, he screens prints onto clay sheets. These he subsequently uses to pressmould, and build sculptural figures. The 'prints' are not precise. Some blockage of the screens results in erratic colour printing, but because of the nature of the work, and the subsequent covering with a variety of glazes, a precise pattern or design is not a necessary prerequisite for the success of the work.

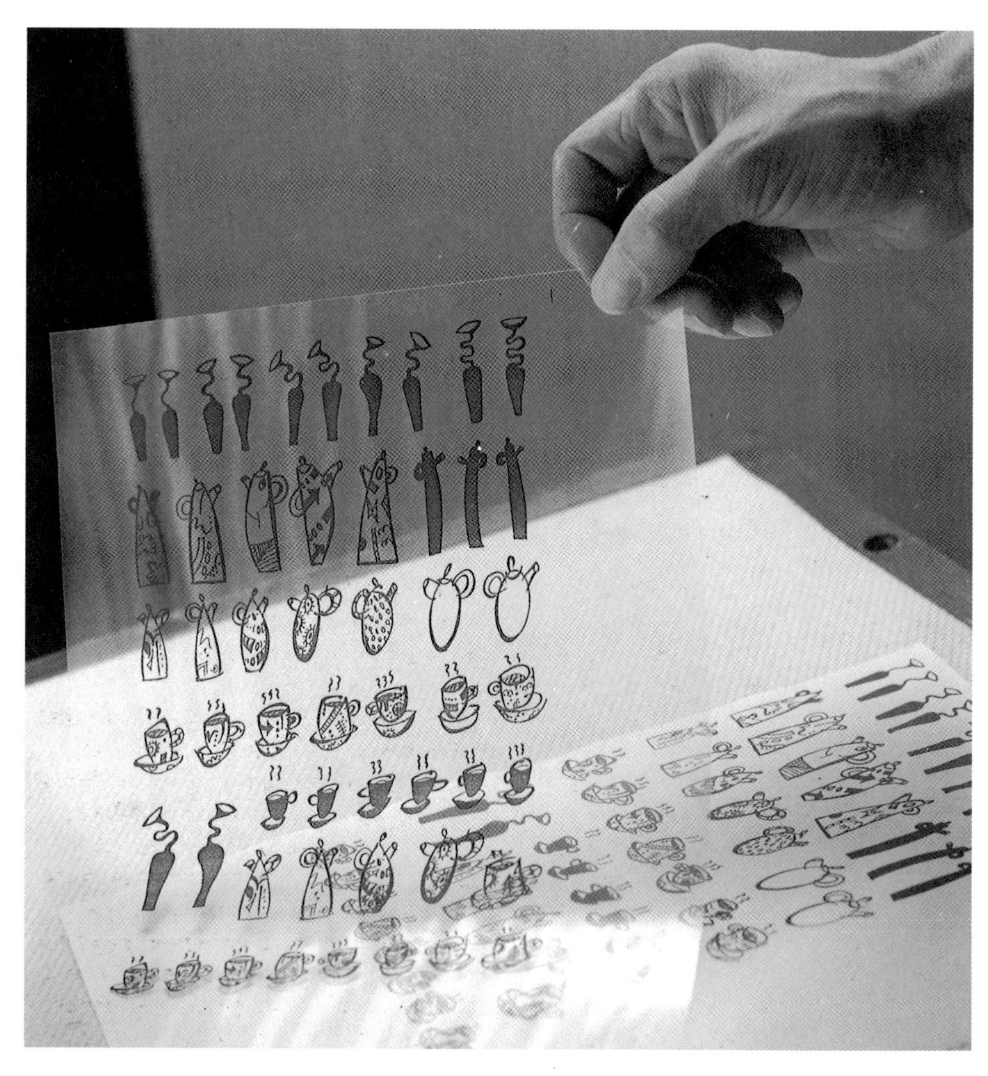

Photocopy transparency with original on table surface. If using photocopy transparencies, reduce exposure time for screen emulsion to UV light. The dark areas of the transparencies are not as dense as prints on Folex Laser Matt. A proper film positive is the densest.

Photosensitive films and emulsion

Almost any design can be transferred to a silk screen by using photosensitive screen emulsions or films. Photo-stencil films are applied to the screen before exposure through a 'positive', to ultra-violet light. Films can be applied from sheets of photo-stencil material, or squeezed on with photosensitive emulsion. When dry, screens are ready for exposure.

The UV source can be as simple as sunshine or daylight, but conventionally, a mercury vapour lamp or carbon arc lamp are used. The type of light source affects the exposure time as does the film type, and density of image on the positive. After exposure, the screen is 'developed' in a spray of lukewarm water, which washes out the unexposed film. Capillex film, and Dirasol emulsion from Sericol,[6] are suitable for this type of screen, and for oil- and water-based media. After use, screens are cleaned in water (or spirit-based cleaners if using an oil-based medium), and are easily reclaimable by using a proprietory screen cleaner.

The 'positive' can be as simple as a found flat object that is opaque or contains areas of opacity. It can be hand-drawn on acetate or tracing film, or a film positive or equivalent.

Making positives for use with light sensitive films

Because a silk screen only prints in colour or no colour, it cannot 'see' tones. A low tech approach to photographic images is to reduce them to simple black and white by photocopying the original, then photocopying the photocopy until all greys have gone. Access to a computer and use of Photoshop software will enable much more sophisticated photographic manipulation of images. In Adobe Photoshop digital images (created by scanning,[7] downloading from the web, or inputting from a digital camera) can be altered in a bewildering number of ways.

Traditionally, screen-printing photographs is done by creating halftone positives. The simplest way to understand halftones is by close examination of a newspaper photograph. Large dots are close together and create the illusion of a dark area, smaller dots are widely spaced and create the illusion of lighter tonal areas. In the same way, a halftone positive breaks up all tonal areas of an original and translates them into a black dot system. Photoshop can do this but allows sophisticated prior image manipulation by altering brightness, contrast, colour balance, and colour levels.[8] In addition, programmes like *Andromeda Series 3 Screens* or *Cutline Filters* (plug-ins) allow an alternative method of producing positives needed to make screens, by creating 'stacastic' images. Instead of reducing and enlarging dot sizes as in a halftone to create areas of light and dark the programme randomly (or in a patterned way) scatters smaller or larger dots corresponding to the tonal areas of the image processed.[9]

A specialist film for use with laser and photocopiers is Folex Matt Laser Film. Although black prints on laser or photocopy transparency films are less dense they will also work as positives.[10] For larger images of higher quality, artwork can be taken to a commercial reprographics bureau who will produce high contrast film positive from paper art work or digital files (as long as they are saved in the correct format). They can also be

[6] Most screen print suppliers will have their own brands of equivalent products.

[7] Even three-dimensional objects can be scanned by most scanners if the depth of the object is not too great.

[8] A detailed and clear description of using the computer for digital image manipulation prior to screen making, and explanations of halftones and mesh sizes, can be found in *Water Based Screen Printing* by Steve Hoskins (A&C Black 2000).

[9] Other digital manipulation programmes including Adobe Illustrator and Andromeda Freehand are suitable for working pattern and image before creating positives for screens.

[10] In addition, ordinary laser printed paper coated with liquid paraffin or vegetable oil renders the paper translucent and suitable for this purpose.

produced in the darkroom, with the right equipment, materials and a basic knowledge of photography.

Printing inks, media

Ceramic inks can be purchased from commercial suppliers. Although they are usually only available in large quantities (5 kg), and as onglaze inks, increasingly studio ceramics companies are supplying suitable materials for printing. Some companies will mix up oil-based inks in smaller quantities (1 kg)

It is quite possible to mix up inks and a number of water-based and oil-based screening media are now readily available from ceramic suppliers. Colour should be mixed with a palette knife or glass muller, in the ratios of 1 part medium to 2 parts ceramic pigment (this can

St Mary Street, Cardiff, Marion Brandis (UK), 1989.Screen printed transfers quarry tiles.

Above: *Roaring Beach*, Ken Ford (Australia) 1998. Mixed media screen print on tiles, paper and backlit transparency. End panels feature 4 colour process on glaze decals, fired to 820°C. Image digitally developed from 6 photographs and over drawing on computer.

vary according to the medium used).

Traditionally, oil-based systems of printing have been used, but these have considerable health and safety implications and the move now is to safer, water-based systems. Oil-based screening media work extremely efficiently, but they involve the printer handling quantities of solvents both in printing and cleaning. These present health and safety hazards in use and storage. If using oil-based systems, good ventilation at all times is necessary, and adequate procedures for safe use, storage and disposal of materials if necessary.

On or overglaze prints

Overglaze colours are most commonly used as inks in decals, because prints are normally applied on top of a fired glazed surface. After application of the decal, the piece then only requires a low temperature firing to mature the onglaze colours.

Overglaze colours have also been developed which allow a four colour printing process as used in the reprographics industry, although they tend to be very expensive.

Tall China Vase, Kosar Kalim (UK) 1999. Decals need not just be screened under- or overglazes; here Kalim uses crystalline glaze, enamel and lustre decal prints on slip cast china.

Canton Lighthouse, Richard Shaw (USA), 1985, height 6.3 cm. Porcelain with overglaze decals. Casting familiar objects and reconstructing them as sculptures, American ceramist Richard Shaw uses multicolour screen printed decals to enhance the illusion of ephemera and everyday objects, that are in fact ceramic.

Above: Close-up showing how inglaze decals can move and distort in firing if slightly over-fired. *From Pots no. 30*, Paul Scott (UK) , porcelain form with decals and lustres. 1200°C, 16 cm (6") tall. *Photograph Andrew Morris.*

Underglaze or inglaze prints

The process can also be used with underglaze colours, applying decals to a pre-fired glazed surface. The actual process of applying transfers allows a remarkable degree of freedom to compose, because the images held within a transparent film glide over the fired ceramic surface and allow for positioning, and repositioning several times. Once complete, the work requires a further firing to just below the glaze maturing temperature, so that the underglaze colour sinks into the glazed surface.

Care must be taken to ensure that the glaze is not overfired, as printed images can distort and move if the glaze becomes too liquid for too long.

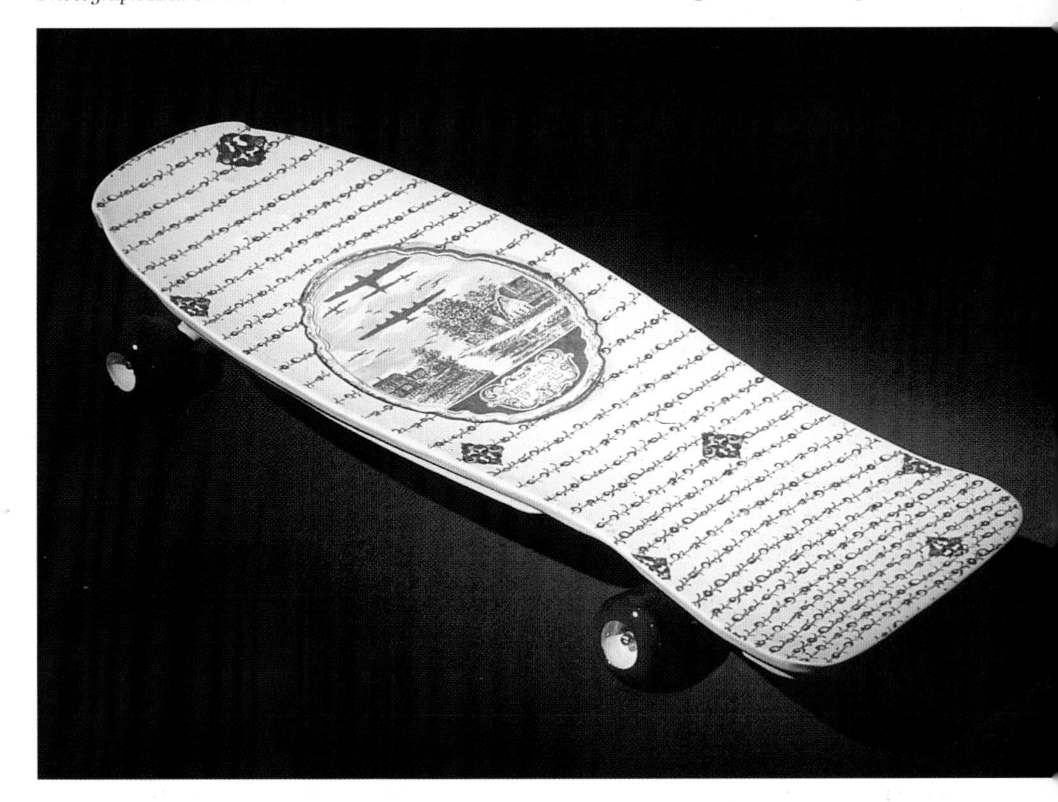

Disasterware, Dresden, 1945 Skateboard Charles Krafft (USA) 1999. Decals on slip cast vitreous porcelain skateboard. *Photo courtesy John Michael Kohler Arts Centre.*

Left: Paul Scott (UK), detail of collage bowls, 1988. Painted underglaze, screen printed inglaze decals, transparent glaze and lustres. *Photograph Andrew Morris.*

New Vision – Teapot #A70723, Les Lawrence (USA). Photo silk-screen monoprint on porcelain, stainless steel, 1995. 30cm x 22cm x 7cm (12"x 8.75"x 2.75"). Les Lawrence screen prints with stained slips onto plaster bats. He works into prints by drawing and painting before removing by pouring casting slip onto the bat. When leatherhard, the slab is lifted from the plaster, a thin sheet of clay including in its surface the screen printed image. He uses these screen printed monoprints to construct his surreal forms, incorporating wire in the building process.

One Hand Must Wash the Other, Kristín Isleifsdóttir (Iceland), 40cm x 40cm (15"x 15"). Printed with black underglaze colour mixed with syrup and liquid soap on 'thin silk paper'. Transferred to bisque fired white stoneware form. Very thin layer of transparent glaze. Sprayed onto the printing after firing 700° C. The letters (words) are the unprinted area. Glaze fired to 1280°C. In the high fire the glaze makes the black underglaze bluish and fluxes the unprinted area (the letters) so they become goldish.

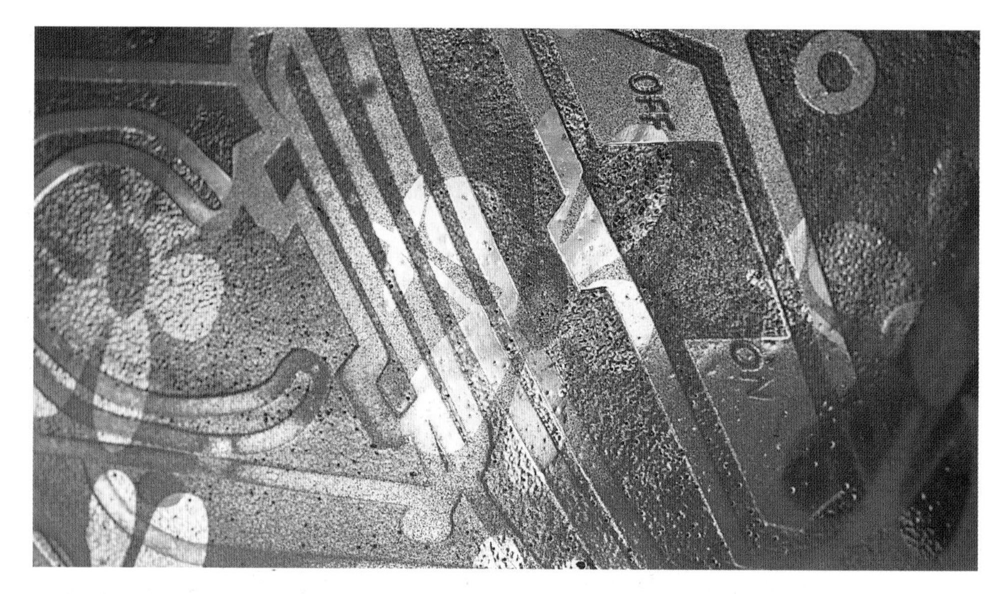

Pifco and flower detail, Katie Bunnell (UK) 1996. Underglaze and lustre on bone china plate 30cm (12") dia., with sand blasted surface. Produced using Adobe Illustrator and Photoshop, with Cutline 3, and a Roland Cam PNC 1100 cutter/plotter machine. Bunnell used a complex combination of processes to explore the potential of CAD/CAM technology in Designer Maker practice for her PhD research.

Other indirect/transfer methods

Uncovercoated prints on decal paper can be transferred face down onto leather-hard clay. By sponging the back of the absorbent gummed paper the print is released from the paper surface and adheres to the clay, although it may need some persuasion by gentle rubbed pressure. This process will work with laser and photocopy prints, oil and some acrylic-based ceramic inks, although some crack as the clay dries.

Direct printing

If printing directly onto clay, then the logical choice of media is a water-based system, because it avoids most of the health and safety hazards of oil-based ones.

Custom made decals

It is possible to have transfers of designs made up on a relatively small scale, thus doing away with the need to go through the whole printing process.

Industrially produced decals

The ceramics printing industry churns out thousands of decals each day, and it is possible to buy some of these as 'open stock' sheets of transfers. Generally purchased in large quantities by industrial producers of tiles, domestic or gift-ware, the selection of images is, in general, limited to the taste of the relevant pottery industry. Blank sheets of colour are also available from some sources.

Above: Ane Katrine von Bülow (Denmark). Porcelain vases with oil printed wire pattern. Printed onto tissue and transferred to porcelain slipcast forms. Matt or shiny glaze, 1280°C. This low tech version of screen transfer printing is based on the tissue transfer system used for copperplate engraving, flexography and monoprinting. It removes the need for covercoating, but is less precise in its transfer of detail as the ink has to be wet or sticky to make an effective print, and some distortion is inevitable as pressure fixes the ink to the ceramic surface. Von Bülow makes tissue stencils specially to fit the forms ('Like a dress made to fit a woman').

Right *Hokusai – this day,* Maria Geszler (Hungary). Screen print on porcelain 97cm x 34cm x 16cm (38"x13"x 6"). Directly printed on porcelain slabs then formed by press moulding in plaster moulds, high fired. Maria Geszler has been screen printing onto porcelain with stained slips and ceramic colours since 1978.

Above: *Genetic Café*, Claudia Clare (UK) 2000. Direct underglaze screen print.

Right: *Inigma* Fleur Harvey (UK). Silkscreen prints in velvet underglazes on stoneware. *Photograph, Alan Hayward.* Harvey prints directly on leatherhard clay with slips and Amaco velvet underglaze colours. 'They have a lovely semi-translucent quality which give the appearance of velour if left unglazed.'

Below: *To Open as Door,* Scott Rench (USA) 1994. Ceramic. direct screen print in Amaco velvet underglaze colours 36cm x 66cm (14"x 26").

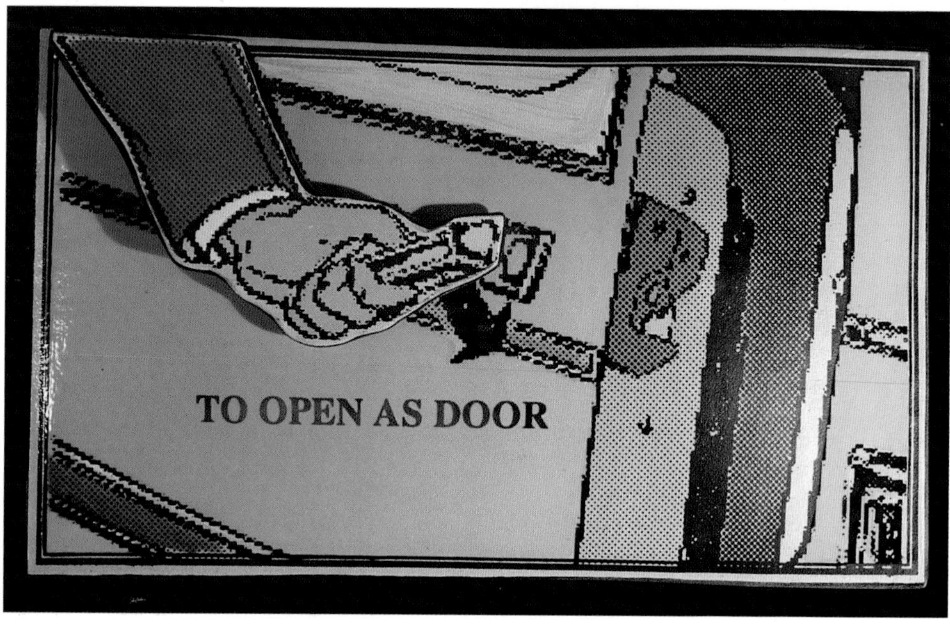

Right: *Blue as the Sky after Rain*, Ursula Smith (UK) 2000. Stoneware with celadon glaze over relief tile 25cm x 25cm (10"x10"). Made by screen-printing shellac on leather-hard clay. Later wiped with wet sponge to erode clay surface where shellac was not present.

Below: *On a Spin*, Claire McLaughlin (Ireland) 28cm (11") dia., direct screen print on clay, jigger jollied to distort image.

Essex Man, Grayson Perry (UK) 1998. Coil pot with sgraffito, sprigged and custom decal surface.

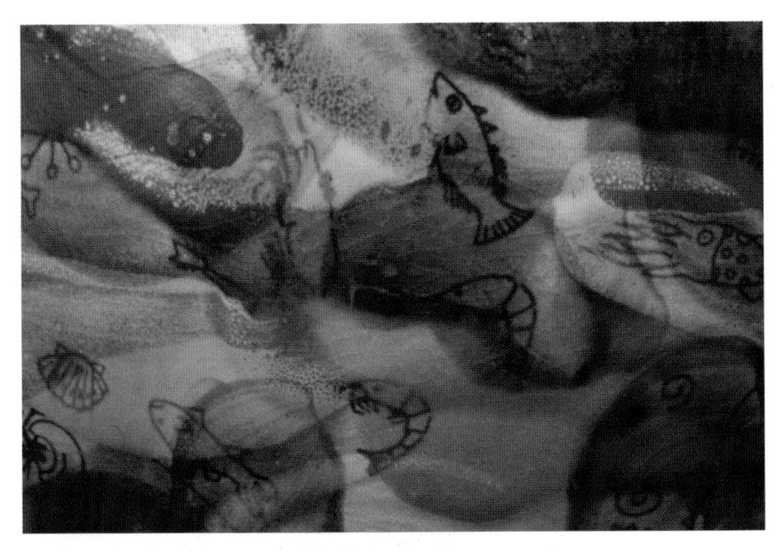

Left: Kate Malone (UK), detail of glazed surface, decals buried under layers of glaze.

Above: Beaker by Åsa Lindström (Sweden). Lindström uses the repetitive potential of print in order to create pattern. As well as having her own studio practice, she does exclusive designs for production by Rorstrand porcelain and also works to commission on large tiled panels for public places.

Above: *So many men, so little time*, Léopold L. Foulem (Canada) 1993-1997. Ceramic and found objects with open stock decals.

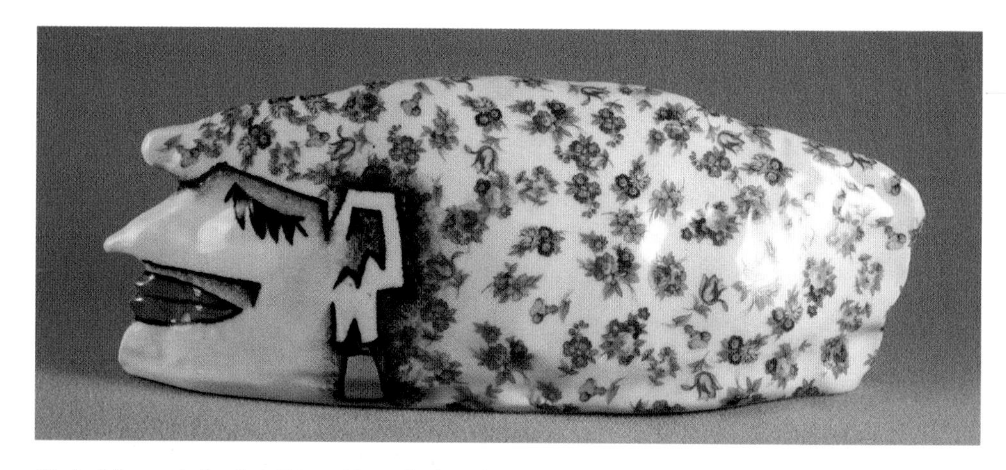

She had flowers in her hair, Rimas Visgirda (USA). Open stock decals, with drawn and painted details on porcelain form, 10cm x 28cm x 9cm (4" x 11" x 3.5").

Scotland, Philomena Prestsell (UK). Open stock decals on slip decorated hand made teapot.

Cups and Saucers, Marianne Hallberg (Sweden). Hand made porcelain cups on ready made saucers, with open stock decals and painted details.

2 Cylinders with Blue, Bodil Manz (Denmark). Porcelain with cut blue decals.

Chapter Six
Lithographic printing

Lithography is a chemical printing process. It relies on the fact that water and grease do not mix. Very simply, a drawing or painting is made with a greasy media onto a specially prepared plate. After treatment with a number of chemicals, the plate is moistened with a damp sponge and then inked up with oil-based inks. The ink only adheres to those areas covered with the greasy media. Upon contact with paper on the litho press, images and designs are transferred from the drawn and printed areas. The amount of ink transferred by this chemical printing process is quite adequate for normal applications, but ceramic prints require a greater density of colour than paper ones.

So, with ceramic litho prints, instead of using ceramic inks, litho varnish is used for each colour. After passing through the litho press, the decal paper is dusted with the desired ceramic colour. It sticks only to those areas printed with varnish. After drying, the decal can be overprinted from another plate, and more colour dusted on. This process can be repeated, but commercially, four-colour prints are usually the maximum. In the ceramics industry, lithographic printing is almost exclusively used with 'predictable' onglaze colours. Because making lithographic decals demands access to specialised and expensive equipment, this sort of printing can realistically only be done in a print workshop. To my knowledge there are still very few ceramic printmakers (outside industry) who have worked in this area in this manner.[1]

There are however, low tech methodologies using Newsprint, photocopies and laserprints that fit into the general principles of lithography. Here it is the repelling action of the ink or fused toner, and absorbing quality of paper that act to make at least two different print processes possible.

Don Santos has developed the Viscosity Transfer process:

> Black and white photocopy machines use a heat process that melts a plastic component in the toner to make it permanent. Thus the black of a photocopy is non-porous. When you wet a photocopy with a solution of water and a small amount of gum you can see the water roll off the printed area and saturate the plain white paper. Ceramic stains or oxides are mixed with linseed oil to make ceramic ink that is sticky. This stickiness or viscosity enables the ink to adhere to the nonporous photocopy ink but not water saturated paper. The ink is applied with a soft brayer or roller, and excess cleaned with the same gum water solution, then blotted dry. The paper is then lifted and placed on a leather hard clay surface, blotted and

[1] Record of University of Tasmania, Hobart research students Deborah Gataric and Megan Alford's work with litho decals can be found in *Ceramics Technical*, No 8, pp. 14-21.

Right: *Birds of a Blue Persuasion,* work in progress, Don Santos (USA) 1999. Viscosity Transfer print on white stoneware with mason stain #6371, 23cm x 26cm x 18cm (9"x 10"x 7").

Below: Removing viscosity transfer photocopy paper after transferring print to clay.

Left: Fired print from photocopy resist. Black lettering (in reverse) on paper has been brushed with water-based ceramic pigment. The printed areas resist the water-based ink and pigment adheres to paper. Transferred to leatherhard clay surface and later fired the finished print clearly shows how the ceramic pigment has been repelled from the original black photocopy toner area of paper. *Photograph, Patrick King.*

Left: Patrick King peeling away paper showing transfer of image to leatherhard clay surface. Here the lettering was originally white with photocopy black back-ground.

Left: Patrick King finished plate awaiting bisque firing.

A World of Difference, Patrick King (UK/Switzerland). Stoneware, 40 cm (16") dia., with photocopy litho printed surfaces.

light pressure applied. The paper is then removed and the image is thus transferred.

Using the same physical properties of paper absorption and toner repellent, an opposite methodology has been developed by Martin Möhwald and Patrick King. For their process the ceramic colour is fixed to the absorbent white paper rather than the toner.

Left: Martin Möhwald (Germany), painting black underglaze onto unprinted white lettering, the black surrounds being relatively non absorbent repels the painted line.

Left: Treated paper is dipped into slip allowed to become leatherhard before application to wetted clay surface.

Below: Removal of paper reveals transfer of print.

Martin Möhwald printed form.

Chapter Seven Direct photographic methods

Direct photographic emulsions are liquid, light sensitive emulsions applied to ceramic surfaces. After drying they are exposed to light and 'developed'. The resulting print is part of the ceramic it was applied to. There are several emulsions applicable to ceramic surfaces that do not require a firing, i.e. the finish is a non-ceramic one. Emulsions which do produce a ceramic colour do so because the emulsion contains a ceramic pigment, or the light sensitive film produces a latent image which is revealed by dusting on a ceramic colour.

Direct emulsions work by allowing light through a colloidal substance to activate light sensitive chemicals. A colloid is usually an organic substance, soluble in water (gelatine, gum arabic, egg) which when mixed with a light sensitive chemical takes on important characteristics. Some colloids are rendered insoluble when exposed to light, and others lose their stickiness.

Direct photographic emulsions using ceramic pigments

The 'Gum Bichromate Process' is an old photographic process, relying not on the light sensitive qualities of silver bromide (as with conventional black and white photography), but on the selective hardening of gum arabic, pigment and ammonium or potassium bichromate when exposed to ultraviolet light. (In the following account, 'bichromate' means the same as 'dichromate'.) All processes, except exposure, are undertaken in subdued light. Paper or ceramic surfaces are coated with a gum bichromate solution, which is exposed to UV light through a negative transparency placed in contact with the emulsion surface. The print is 'developed' in water, which washes away the gum and pigment from the unhardened, highlighted areas. In the case of ceramic 'prints' the colloid contains ceramic pigments in the mixture.

Jim Bennett has been using this process to make ceramic photographs on tiles. The following is his basic recipe for an emulsion:

1 part ceramic pigment,
2 parts water,
2 parts gum,
2 parts ammonium dichromate.

He uses onglaze or enamel colours, gum arabic or office glue (gloy), or even whole egg (for very porous surfaces like bisque) as the colloid. The mixed emulsion is painted onto a fired, glazed tile surface with a soft paintbrush. After drying, the tile is exposed through a contact negative transparency to a UV light for up to 10 minutes. The printed image becomes clearly visible during exposure if the negative is lifted carefully at a corner.

Small photo-ceramic tile by Jim Bennett (UK) and Claire Mc Laughlin (Ire) 1993. Vitrified stoneware with ceramic enamel powder, printed by gum bichromate.

The 'printed' colour washes out slightly on developing, so first experimentations might well use judgement by eye as a guide to exposure time. The emulsion is developed by gentle brushing of the surface in cold water. Further coats of emulsion containing different pigments may be subsequently applied, exposed and developed. He suggests using a pale colour for the first coating, with subsequent pigments becoming darker, and having shorter exposure times.

The process is one where no hard and fast rules apply, where experimentation can lead to subtle and delicate photographic qualities. The variants are the density of negatives, type of light source (prints can be exposed using daylight instead of a UV lamp, but exposure times will be longer), length of exposure, strength of bichromate solution, colour (different colours may require different exposure times), glaze and firing temperature.

In a book *Alternative Photographic Processes*, Kent E. Wade documents the printing of large tiled panels using bichromate. This involved the use of a commercially available screen emulsion made from a 'polyvinyl-alcohol/ polyvinyl-acetate solution' as the colloid into which pigment and bichromate was added. He used 'up to a maximum of approximately 50% dry glaze ingredients . . . without affecting its light sensitive qualities'. He observed that even without the addition of ceramic pigments to the emulsion, the dichromate 'when kiln fired to varying temperatures and under differing conditions, will exhibit a wide range of colours . . . the result of the dichromate's turning to chrome oxide'.

Ammonium and potassium dichromate are available from photographic suppliers. Ammonium salts are more light sensitive, and require less exposure time than potassium salts. They are

Photographic ceramic print on plate by W. Weerasuriya (Sri Lanka).

potentially dangerous chemicals, and proper care should be exercised when using them.[1]

A potential difficulty with ceramic pigments is that they are particularly dense, and may obstruct the passage of light through the emulsion, with the result that the gum may set on the surface but not underneath. Jim Bennett has not found this to be a problem using onglaze enamels, but it might be one if colloidal emulsion was painted on too thickly, or if pigmentation of the colloid was dense (underglaze colours might need a denser application). One way of avoiding this would be to apply the colloid as a thin emulsion, expose and develop, repeating the process to progressively build up a thicker emulsion base.

An alternative might be to try the 'Carbon Transfer Process'. In this case, the bichromated gum solution with pigment is painted onto tissue paper. The tissue paper and emulsion is exposed emulsion side up to UV light through a negative transparency. Then the paper is stuck, emulsion side down, onto a ceramic surface. The print is washed in warm water, removing the tissue, and develop-

[1] See Health and Safety section.

Above: Thomas Sipavicius dusting ceramic pigment onto exposed gum bichromate print.

Below: Thomas Sipavicius washing plate after dust application.

Who am I the only one? 1-10 tile no 3, Thomas Sipavicius (Lithuania/Hungary). Fired ceramic gum bichromate print, 30cm x 30cm (12"x12"). Sipavicius researched ceramic gum bichromate processes at the International Ceramic Studios in Hungary during 1998 and 1999.

ing the emulsion by washing off unexposed areas. The hardened pigmented gelatine adheres to the ceramic surface in a depth proportional to the amount of light falling on it through the negative. Thus the deep areas appear dark and the shallow areas light, with perfect half tones.

An alternative, particularly ceramic adaption is the use of bichromate with gelatine and a sticky substance such as honey, glucose or sugar. In this case the solution is applied to a ceramic surface and allowed to dry. It is then exposed with UV light through a positive transparency. Where the UV light strikes the film, it modifies the surface of the emulsion so that it is no longer sticky, but this is a gradual process, and the bichromated gelatine remains become less sticky in proportion to the amount of light falling on its surface. After exposure, the latent image is revealed by dusting the surface with a ceramic pigment. The most colour sticks to those areas least exposed to the light. As with other 'dusting on' processes this should be done in an area with adequate exhaust ventilation.

Of these methods, dusting on was perhaps the first to be used in making ceramic photographs and Thomas Sipavicius has perfected a methodology for contemporary use.

Peter Frederick, a photographer who has worked extensively with gum bichromate processes, has said: 'Gum printing is not a complex affair ... It's simple, little hardware is needed (not even a darkroom or enlarger) and allows your creativity full reign..... But you do need to undertake each step in the process with care and patience.'

Non-firing photographic emulsions

Liquid Light and similar emulsions are made from more conventional photographic ingredients, and as such require (albeit makeshift) darkroom conditions for use. The emulsions cannot be fired, and so any use of them must be after all firing processes have been completed.

The instructions for their use stress that surfaces onto which the emulsion is to be applied must be properly prepared. The recommended preparation for ceramic surfaces is a painted layer of oil-based primer paint, or polyurethane varnish (if necessary, thinned with white spirit).

At room temperatures the emulsions are solid gels, and to render them usable containers are placed in hot water. Once liquid, the emulsion can be applied by

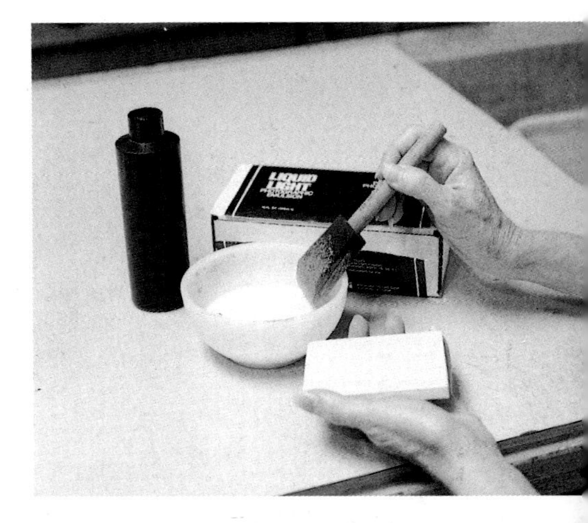

Liquid Light applied to title with sponge applicator. Image is exposed and tile is placed in developer, where image appears on tile. After fixing in two baths the tile is rinsed in running water. *Photographs by Linda McRae.*

painting, rollering, sponging or spraying (with adequate ventilation). Like paint, too thin a layer of emulsion will show streaks and brush strokes, so if these are not desired then multiple layers must be applied allowing the emulsion surface to almost dry before re-application.

Small test tiles should be coated at the same time as the main surfaces: these are

Death of the Virgin, Linda McRae (USA) 1990. Liquid Light photo emulsion on smoked porcelain. 50cm x 40cm x 8cm.

used to test the correct exposure. All coated surfaces should be dry before exposure and drying is best done in complete darkness.

Exposure is by contact negative or enlargement, or by projecting a black and white or colour negative. Full instructions are supplied with the emulsions. Upon developing in conventional photographic developer, the emulsion is washed for 30 minutes in running water to make the image archivally permanent.

Liquid Light, or similar emulsions like Silverprint Emulsion, are different to many other processes described in this book in that once applied, exposed and developed, the emulsions produce archivally perrnanent images on a ceramic surface without firing. Because of this it is probably one of the processes least likely to appeal to the 'traditionalist', who will see this as yet another denial of a fundamental ceramic process: fired colour or quality. Yet the work of Linda McRae shows how sensitive, complex and successful pieces of work can be made using this technique, combining smoke-fired ceramics with photographic qualities produced with Liquid Light.

Some time ago, enquiries about her smoke-fired ceramics revealed that some observers thought that there was some photographic process involved in the pieces because of some of the qualities of tone achieved by the smoking.[2] Although there wasn't at the time, it provided the germ of an idea that led to experimentation with a number of photographic emulsions. Settling on Liquid Light, she uses it on sheets of biscuit-fired and smoked porcelain.

Instructions reveal that preparation of the ceramic surface is important for the permanence and stability of the finished emulsion. Linda has used water-based paint or acrylics to prime the biscuit, finding that although they do not adhere well, they do allow manipulation of the exposed image through drawing, stretching and peeling. The use of transparent layers of polyurethane varnish in contrast, allow the colour and texture of clay to show through the emulsion, and produce a more permanent and reliable surface. By careful planning, the photographic image can be integrated into a composition making use of a variety of ceramic techniques and firing methods.

As with other photographic emulsions, there are a number of variable factors which affect the quality of printed image: pre-coating, thickness of emulsion, method of application, age of emulsion (Liquid Light has a shelf life and changes as it ages), length of exposure, and, of course, the surface onto which the emulsion is applied.

Liquid Light and Silverprint Emulsions 'contain no hazardous chemicals and require no special precautions'.[3] They do require developing with conventional process chemicals though and due care and attention should be paid to their use, especially if working with large ceramic objects.

Like many of the processes described in this book, photographic emulsions do not to have many applications for 'commercial' or mass-produced ceramic objects, but they do provide intriguing possibilities for those interested in the creative use of light, photography, printing and ceramics.

[2] See also Dick Lehman's kiln prints in Chapter 5.

[3] Silverprint Catalogue 1993, Rockland Colloid Corp. material instructions.

Conclusion… a place for ceramics and print?

Clearly contemporary practice in Ceramics and Print is diverse, but it is possible to identify work as being created within one or more of four categories:

Work produced by artists exploring the creative printmaking potential of ceramic materials, in the traditions of fine art printmaking and explorations of printed (ceramic) surface by contemporary ceramists, typified by the work of Mitch Lyons (USA)[1] monoprinting off a large china clay slab, onto paper with slip mixed with pigments as the ink, or Les Lawrence (USA)[2] whose silk screen monoprints are taken off plaster, producing graphic sheets of porcelain for use in constructing ceramic forms.

Work produced by artists referencing, subverting and appropriating diverse ceramic, and other traditions. Work may be personal, political, or related to perceptions of art, and involve an extensive group typified by artists as diverse as Conrad Atkinson (UK/USA), Leopold Foulem (Canada), Robert Dawson (UK).

Production studio pottery using and developing traditional ceramic print methodologies (e.g. sponge, rubber stamping and decals).[3] Typified by potters and ceramists, who use print as a purely decorative adjunct to the surface.

Industrial production and innovation. Of interest here because of potential for creative appropriation. Typified by Spode, (digital development of engraved images, for screen printing).

In identifying the differing strands of contemporary practice it is tempting to add labels or classifications, but the promiscuous and fluid nature of much engagement in the field makes an effective definition by label untenable, and undesirable. There is also room to identify subgroups within the four main areas, for example a differentiation between those operating originally from printmaking, and those originally from ceramics. Problematic here though is the tendency for younger artists to be unattached to any particular discipline. In addition, the identified groups are not actually rigid or fixed, as contemporary artists move between them, or operate in one or more of them at any on time.

Whilst industrialists, and production potters, are unlikely to engage with the creative explorations of those operating within the first two groups, the converse cannot be assumed. Indeed artists constantly seek to engage with methodologies and in situations outside their traditional studios.

[1] See: www.rosenthalgroup.net/artweb/mitch
[2] See: www.printandclay.net and www.artcumbria.org/hot_off_the_press.htm

[3] Decals: images/designs printed onto special pre-gummed paper with onglaze or underglaze colours in oil medium. Thermoplastic covercoat lacquer subsequently applied. When dry, covercoat forms plastic film containing printed image. This film slides off in warm water (hence 'decal' meaning 'transfer'). Pre-printed open stock decals available from ceramic printers.

Evaluation of printed ceramics depends to a large extent on the context in which the work is made and in which it is exhibited. The printed ceramic work of Jeff Koons, Cindy Sherman[4] and Conrad Atkinson for example are non-controversial interventions by leading protagonists in the contemporary arts.

For those whose work does come from a craft base, for example the potters decorating forms with printed pattern, there is, in spite of Leach's assertion that 'well painted pots have a beauty of expression greater than pottery decorated with engraved transfers, stencils or rubber stamps',[5] little problem today in acknowledging print's value and authenticity as a valid contemporary practice. Similarly for printmakers using paper-clay, and ceramic printmaking methodologies based on industrial pedigree, there is little problem in justifying the work, or finding a place for it in the realms of print exhibitions and print studios. Using a primarily flat picture plane, working within defined and well-documented practice, it basically fits fine art printmaking, which has just a few material adjustments to make.

Issues do arise when the ceramic and print involvements originate from artists whose primary practice is ceramic based. Because writers and historians on ceramics make great emphasis on form and function, work that appears to operate outside the accepted values of the medium, or ignores the two behemoths, is seen as superficial.

Their works, more often than not, are made as a result of a long engagement with ceramic material, and rely on its suitability as graphic medium: Kevin Petrie writes: 'My pleasure in transfer printing derives from the qualities of surface which are unique to the process. Only through onglaze transfer printing can the brightly coloured glossy imagery be achieved.'[6] They reference unacknowledged, but existing ceramic and print traditions and practice, and wider practices and ideas from the visual arts.

Because these and others like them, operate primarily from ceramics (outside traditional printmaking) and are not always promiscuous enough in their contemporary practice to be placed in cutting edge contemporary arts, there is a danger that their work exists in an 'aesthetic no man's land'.

That this ephemeral, unrecognised position is becoming so populated might indicate that those occupying it have something worthwhile to show and contribute. For some the promiscuity is dangerous, a threat to traditional practice and carefully protected academic positions. For others this interdisciplinary mix of tradition and innovation, tests those labels that define work by its material content rather than its intent or location.

Since 1994 when *Ceramics and Print* was first published, things have changed a great deal, yet there is still much more to discover...

---

[4] See *Already Made* Museum voor Hedendaasgse Kunst, Het Kruithuis, s'-Hertogenbosch, Netherlands 1993.

[5] Leach Bernard *A Potter's Book*, Faber, 1940.

[6] Kevin Petrie 1998.

Glossary of terms, techniques and materials

Bite Action of acid or other corrosive agents on metal.

Bleeding Seepage of colour or medium from a printed line. Cobalt frequently bleeds into glaze upon glaze firing.

Colloid Unsensitised compound (PVC, gum arabic or gelatine) used in manufacture of photosensitive medium. Colloidal particles are those of finely divided substance when dispersed throughout another substance.

Colouring oxides Metal oxides which give colour in fired ceramic form. Used in all ceramic colours (onglaze and underglaze).

Copperplate oil Oil medium made from boiled linseed oil, used in making etching and engraving inks.

Covercoat Varnish or lacquer of plastic substance in solvent. Use only with good ventilation. Applied in decal paper through a silk screen, usually over printed image. When dry, forms thin sheet of plastic film which slides off the paper in warm water, taking printed image with it. Applied to glazed or polished surface of fired ceramic, upon firing plastic film burns away (ensure good ventilation) leaving ceramic image in or on glaze or ceramic surface.

Decal paper Paper with gum coating used in production of ceramic transfers. Images printed onto surface with onglaze or underglaze colours in oil medium. Covercoat applied. When dry, covercoat forms plastic film containing printed image. This film slides off in warm water (hence 'decal' meaning 'transfer'). Some decal papers can be purchased pre-covercoated, for specific methodologies (e.g. water-based systems, or laserprints).

Diazo Biodegradable photosensitive chemical used in photo emulsions.

Dry point Very basic intaglio method of producing a line by scratching into a plate with a sharp point. Method capable of producing very subtle qualities in printed image. With metal plates (copper, zinc, etc.), hard steel or diamond points are used. Other materials may be suitable for use with dry point – perspex, plastic, plaster.

Embossed print Print with three dimensional surface created by pressing clay or paper clay into lowered areas of lino plate, wood block, plaster block. All direct intaglio prints are embossed.

Engraving Cutting a design into metal or end grain of hard wood (e.g. box or holly) with a graver or burin.

Etching Creating designs in metal plates, using the corrosive action of acids. Other surfaces can be etched lino by caustic soda, styrofoam and some plastics by the use of lacquers or solvents.

Etching grounds (Acid) resisting coating used on surfaces to be etched. Designs are drawn into the grounds exposing the (metal) surface, which is then etched.

Fat oil Printing and painting medium made by reducing pure turpentine over heat. (In this process toxic fumes are emitted. Fat oil should always be purchased ready made.)

Feathering Using a feather or soft brush to gently remove bubbles from metal plates caused by the action of acid during etching.

Film positive Positive transparency used with light sensitive emulsions (screen printing, nylo print etc.).

Firing The process by which heat turns 'clay' into 'pot'. Clay disintegrates in water, but on firing to at least 600°C, clay turns into a stone-like material which will not break down in water. High firing can render clay impervious to water (at vitrification).

Flux Substance added to a material to enable it to fuse at a lower temperature.

Foul bite Unintentional etching of an area on a plate.

Frit Glaze ingredients already fired to make a stable substance and make toxic materials safer.

Glaze Thin layer of glass fused in place on clay body. A glaze (a mixture of glass-forming substances, fluxes and stabilisers) is applied in powder form, usually suspended in water. On application to biscuited surface, water is absorbed by porous body, and powder remains on bisque surface. On subsequent firing, powder melts, and forms glassy layer. Glazes can be transparent, opaque, translucent, coloured.

Greenware Unfired clayware.

Groundlay oil Sticky oil used for printing, later to be dusted with ceramic colour.

Halftone Most printing processes (apart from direct photo emulsions) cannot 'see' tone. Halftones are created in silk screen printing for example, by breaking all the tones in the art work, and recreating them in a black dot system. The most obvious example in common use is a newspaper photograph. Close examination will reveal an image made up of thousands of tiny dots. Adobe Photoshop will easily produce halftone images for use with silk screens and flexographic plates.

Hardening on Process of firing tissue transfer decorated ceramics to 680°C to 700°C to burn out oil medium and fix colour to biscuited surface.

Ink Medium carrying colour used for printing. Ceramic inks need relatively high proportion of pigment to medium. In screen printing this is at least two parts colour to one part medium. Ready made ceramic inks are not easily available in small quantities, but can be made relatively easily using a glass muller or palette knife mixing powdered oxides, underglaze or onglaze colours and medium very thoroughly.

Intaglio printing Printing from lines, grooves, crevices drawn, carved, engraved or etched into a plate.

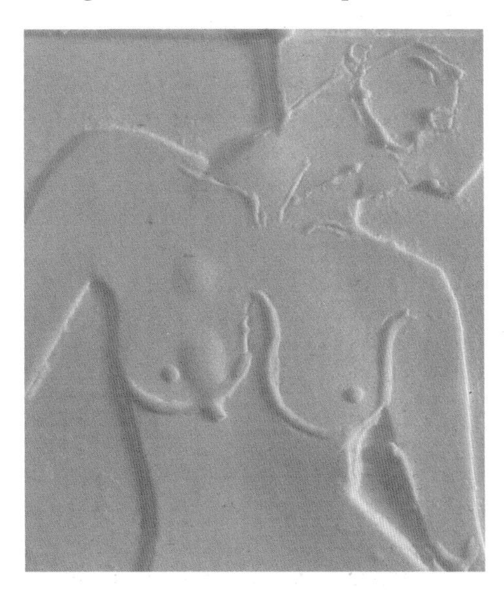

Detail of an intaglio print, *State of Play*, by Mo Jupp (UK). Backlit porcelain print from plaster, 1260°C.

Laser printer High quality printer for computer, fusing printed toner with heat to paper surface. Some toners contain iron oxide and thus fire a ceramic colour (sepia).

Leatherhard Partially dried clay. Body has greater strength than dry clay.

Lift off or snap Gap between silk screen and printing bed so that screen mesh is progressively released from printed surface (clay, ceramic or paper) as the print is made.

Linocut Block print cut into linoleum, a flooring material made from canvas with thick coating or oxidised linseed oil.

Liquid Light Liquid photographic emulsion made by Rockland Colloid in the USA. Other similar emulsions are available including Silverprint Emulsion. Exposed and developed using conventional photographic techniques. Not designed for firing, but for use on ready-fired surfaces. 'Archivally permanent'.

Lithography Invented/discovered by Aloys Senefelder in 1798. A chemical process, based on the principle that grease and water do not mix. Image drawn or painted on specially prepared limestone, or zinc/aluminium plate with greasy crayon or ink. Plate is then desensitised with specially prepared mixture of gum arabic and small quality of acid. Water is then squeezed onto the plate with a few drops of turpentine, and surface wiped clean. Image seems to disappear but is locked into plate surface by grease. Plate inked up by wetting surface with a sponge, then rolling oil-based ink over surface. This sticks to the greasy drawn image, and the rest of the plate rejects the oil. Image can now be printed from the plate.

Lustres Thin metallic coatings on fired ceramic surfaces, usually, but not always, on glazes. Lustres are applied in liquid or semi-liquid form. Lustres are made by dissolving precious metals in hydrochloric acid, the resulting chloride is added to resins, oils and solvents, ready for painting. On firing, the medium and resin burn out creating a reducing atmosphere which produces the thin layer of pure metal. Ensure good ventilation when using and firing lustres.

Mesh count Number of threads per inch or per centimetre in silk screen fabric.

Monoprint or monotype Print in an edition of one. Printed painting or drawing. Image drawn or painted on glass, plastic, metal, formica, board, plaster, print taken by covering area with clay, paperclay, or paper and rubbing, or exerting pressure (e.g. from rolling pin or etching press). Another type of monoprint is made by inking up glass or similar plate with ink of stiff consistency. Pottery tissue paper (or similar equivalent) placed onto inked area and image/design drawn onto back. Where drawn line exerts pressure on plate, ink is picked up. Removal of tissue reveals design in reverse where paper has been in contact with ink. Paper then placed (inked side down) in contact with clay or biscuited surface and rubbed down. Image transfers to ceramic surface.

Offset printing Indirect printing by depositing an image from a roller or plate to another or from a silk screen to a plate, and then printing it on paper or clay surface. Most commercially-produced lithographic prints are produced in this way.

Overglaze colour (onglaze, enamel) Ceramic colours commercially prepared by fritting and grinding, designed to be used on top of glazed surface. The colours normally fuse onto the glazed surface between 750°C and 850°C. A wide range of colours, but because they are on top of the glaze they are less hardy than underglazes or colouring oxides, and more liable to be damaged by abrasion or chemical attack.

Open stock Ready-made decals available from industrial ceramic printers. Individual sheets of transfers can be purchased for use on ceramic surfaces.

Paper clay Compound of paper pulp and clay. As a general guide, percentages by volume from 25% to 50% of paper pulp in the clay body. For printmaking use, a mix of 33% paper pulp to 67% porcelain slip produces a strong paper-like material suitable for printing on. Different paper pulps produce differing effects upon firing. White blotting paper works well, burning out to produce a pure porcelain body.

Photo emulsions Light sensitive liquids, used in direct photographic prints onto ceramic surfaces, or in screen printing.

Photopolymer or **Flexographic** plates: water washable light sensitive printing plates. They consist of removable protective film, a light sensitive layer, an anti-halation layer and a steel support. The protective film is transparent and offers no protection against unwanted light but is useful for handling before exposure. The light sensitive layer is the main body of the plate and sensitive to both light and water before processing; the anti-halation layer provides protection against reflected light and the metal support is useful in inking and printing. Available in a variety of thicknesses for different applications. Also available in solvent washable form.

Plaster of Paris Fine white powder of heated gypsum. Plaster bats and moulds are water absorbent, and used traditionally for slip casting forms. Of particular use to the ceramic printer for monoprinting from, offsetting and intaglio work.

Pottery tissue Thin tissue paper used for copperplate engraving transfers. Also usable for transferring designs printed in other ways (screen prints, lino prints, etchings, monoprints). Alternatives include large sheets of cigarette paper, or other fine tissue.

Relief printing (block printing) Printing from a design standing in relief. Raised areas generally receive a coating of ink from a roller, or in the case of rubber/sponge stamps from colour on a pad or tile. Printing takes place when inked surface comes in contact with paper or ceramic surface.

Rollers Made of rubber, gelatine or plastic. Used to roll printing inks on plates, stones, blocks etc.

Rubber stamps Block printing method employing moulded or cut rubber. Usually applied to biscuit ware where porous body absorbs media quickly, but also applied to glazed ware. In industry mainly used for back stamping, but also used to stamp gold. Stamps are made from brass engravings, or etched polymers (cheaper, but not as high a quality). Can also be made from carved and cut rubber erasers.

Scanner Facilitates input of images into a computer.

Serigraphy or screen printing Method of printing where ink is forced through a mesh stretched on a frame. Screens can be made from silk, organdie, multifilament polyesters, stainless steel, nylon, and monofilament polyester which is probably the most suitable for small-scale ceramic use. Where mesh

Contemporary sponge decorated mugs by Moorland Pottery, Nicholas Mosse Pottery, Bridgewater pottery and jug by Brixton pottery. *Photograph by Andrew Morris*

is blocked by paper, photographic film, or other media no ink can pass through. Widely used in the ceramics industry for direct printing and the creation of ceramic transfers. A process easily adaptable for small-scale use.

Slip Suspension of clay in water. Used for casting, and decorating clay.

Soft soap Potash soap made from vegetable oils. Used as lubricant in transferring designs from pottery tissue to ceramic surfaces. Also used as a size on tissue when printing copperplate engravings and etched plates.

Sponge printing Block or relief printing method where colour in a water-based media is held in the pores of a sponge (natural or synthetic). Designs cut in the sponge transfer to clay or biscuit surface upon contact.

Squeegee Rubber blade set in handle used to force pigment through silk screen mesh.

Turpentine Distilled pine resin. Used in lithography, and as an extender for fat oil.

Turps substitute or white spirit Petroleum distillate, used for cleaning oil-based inks and media.

Underglaze colour Commercially prepared ceramic colours (from metal oxides) designed to be used under glazes. Available in powder or (semi)liquid form. Contain small amount of flux to help adhesion to clay or biscuit body. Glaze or body stains are prepared ceramic colours without the flux, and are therefore slightly stronger in effect.

Wax resist Wax used as a masking medium where no glaze or ceramic colour is required. May be screen printed.

Health and safety

You should be aware of the potential environmental and health and safety implications of any creative processes you undertake. Wherever possible this book has encouraged the use of the least toxic materials possible for any particular process. However, it is possible for individuals using print and ceramics to come in contact with a number of toxic substances, indeed a combination of hazards common for the printmaker and the ceramist.

The main routes of exposure to harmful chemicals are:

Inhalation The main route of entry for airborne chemicals and dusts. These can directly affect respiratory organs, but also through absorption into the bloodstream, the blood, bone, heart, brain and all other organs.

Avoid inhalation Use non- or less hazardous chemicals. Ensure good ventilation, especially when using solvent-based products – oils, media, covercoat, cleaning solvents. Reduce dry ceramic powder processes to a minimum ensuring they are done with good local exhaust, and wearing suitable respirator or face mask. Ensure good efficient cleaning system is employed in the studio, minimise dust production and circulation.

Ingestion Swallowing chemicals is not normal practice, but ingestion by accident is quite possible if care is not taken. Accidental ingestion by eating food, drinking and smoking in the studio or with dirty hands is the most likely. In addition, if chemicals have been inhaled, the lungs' natural cleaning system means that these may be expelled by the lung where they can be swallowed.

Avoid ingestion Do not put contaminated hands near mouths or eyes. Do not contaminate hands directly if possible (see below). Do not eat drink or smoke in the workshop. Clean hands thoroughly after working.

Skin contact Chemicals can be absorbed through the skin, entering the bloodstream as above. In particular, some chemicals cause skin allergies and irritation. Metal oxides can be absorbed through the skin with the danger of long-term cumulative effects. UV light is known to cause skin cancer, other substances can cause ulcerations or burns.

Avoid skin contact Use the least toxic chemicals. Wear protective clothing (goggles when using acids), gloves, barrier cream. Dress any cuts or wounds properly and ensure they remain uncontaminated. Clean hands, and particularly under fingernails thoroughly after working. Maintain high standards of hygiene in workshop.

Injection Unlikely to be a major hazard in a ceramics or print studio, but ensure that sharp objects are cleared up and not hidden under sheets of paper, cloth or polythene.

The Control of Substances Hazardous to Health Regulations 1988 (COSHH) in the UK stipulate that every employer (including the self-employed) who uses, or whose employees use, hazardous substances must assess their dangers, the precautions that are being taken and areas where control needs to be improved. 'Hazardous substances' include typing correction fluid, aerosols and glues, so practically every workplace is obliged to do at least a brief assessment. The results of your assessment must be recorded and may be examined by a Factory Inspector.

In order to make your environmental and health and safety assessments you need to be aware of the potential dangers of any process you undertake, and the properties of any materials you use. In the UK when ordering materials you are legally entitled to request Health and Safety Data information sheets (or MSDS reports in USA). These will detail any hazardous substance contained in the materials you have purchased, recommend methods of storage, use and disposal.

Levels of toxicity

A poison is a 'substance which when introduced into or absorbed into an organism destroys life or injures health' (*Concise Oxford Dictionary*). In reality, many substances regularly ingested in small quantities (alcohol, salt, caffeine etc.) would be toxic if ingested in massive quantities. The human body has a variety of defence mechanisms against toxic materials, but if these are overloaded or fail for some reason then poisoning of some description is the result. Different people have different tolerances to poisons. An exposure level that may not have any perceivable effect on one person may be debilitating for another.

Nowadays there are defined Occupational Exposure Standards (OES) or Limits (OEL) over which a substance ceases to be regarded as 'safe'. OESs for gases, vapours and dusts are measured in parts per million parts of air (PPM), or milligrams per cubic metre of air (mg/m3). Two criteria are used in expressing exposure limits, quantity and time. Time is quantified by Long Term Exposure Limits (LTEL) (generally eight hours working at a time, in a 40 hour week), and Short Term Exposure Limits (STEL) (which has a reference level of ten minutes).

If there is clear evidence of harm resulting from higher levels of exposure, the OES is given a Maximum Exposure Limit (MEL). This is also expressed with a LTEL and a STEL. (If you find you are working with materials with MEL then stop using them and find alternative products.)

When obtaining Health and Safety Data Sheets from manufacturers, the seemingly bewildering figures and chemical definitions are related to these limits. They are produced for industrial use, and are assumed to be the exposure safe for 'healthy men'; not children, women (especially when pregnant), the old, or people with illnesses, and it is difficult for the individual to monitor these figures. A more common way of realising that you are using hazardous substances is by product labelling. Extremely hazardous products are labelled 'Very Toxic', less hazardous 'Toxic', below that come 'Harmful', 'Corrosive' and 'Irritant'.

Children

Children should not be exposed to toxic substances. They have a more rapid metabolism than adults, and so they are more likely to absorb poisons. They have smaller lungs and a smaller body weight, so safe levels for adults cannot be said to be safe for children. Any children working in the area of print and ceramics should be using non-toxic substances and processes (and this is quite possible). Your own children should not have access to your workshop if you are using any toxic substances.

Pregnant women

Risks to the unborn child are high and so pregnant or nursing mothers should avoid any exposure to toxic substances, solvents in particular.

Hazardous substances in ceramics and print

Although environmentally responsible and friendly methodologies in printmaking are evolving rapidly, you may yet encounter a number of hazardous substances if you take up printmaking seriously. Specialist publications deal with this area (see Bibliography). Particular hazardous materials referred to in this book include:

Acetic acid Used in photography for stop baths. In concentrated form is irritant, corrosive and flammable. Vinegar is dilute acetic acid (about 5% acetic acid). Mildly irritating to skin and eyes. Can be used in water-based screen media for ceramic inks.

Aerosol car paints Used to make aquatints. Contain solvents. Avoid breathing mists, ensure good ventilation when using. Aerosol paints can be explosive in mist form, avoid sparks or other sources of ignition.

Aerosol lacquers: Can be used as a substitute for covercoat. As above.

Ammonium dichromate (ammonium bichromate) Used in light sensitive emulsions. Skin contact can cause allergies, irritation and ulceration. Chronic exposure can cause respiratory problems. Suspected carcinogen. Wear protective clothing, gloves and goggles. Explosive in dry state, contact with combustible materials may cause fire.

Caustic soda (sodium hydroxide) Corrosive, can cause severe burns, and respiratory irritation. Used to etch lino, also used in some screen cleaning preparations.

Clay Contains 'free' silica, hazardous in dust form. Long-term exposure to clay dusts can cause lung scarring disease called 'silicosis'. Ensure good workshop practice, eliminating dry clay processes if possible. Wet clean all working surfaces after use.

Covercoat Contains trimethyl benzene. Low order of toxicity, but irritant. High vapour concentrations can irritate nose and throat.Prolonged skin contact can cause dermatitis. Ensure good exhaust ventilation when using. Toxic emissions on firing.

Gum arabic Mildly toxic, can irritate the skin and respiratory system and cause sensitisation. Ensure good ventilation if danger of dusts or mist forming. Safest in liquid form, but shorter shelf life.

Lead Toxic, cumulative poison. Use of raw lead banned in potteries. Lead compounds are now used in safer, fritted forms. Ensure good ventilation and workshop practice when mixing glazes. Fired pottery with lead glazes can be dangerous if underfired, or in combination with copper: Poisons can be dissolved with acids in foodstuffs (vinegar, juice etc.).

Liquid grounds A mixture of asphaltum, beeswax, rosin and a solvent. Check what the solvent is. Ensure it is not benzene or chloroform, both carcinogens.

Lustres Contain a variety of chlorinated and aromatic hydrocarbons. Always ensure good ventilation when working, and especially on firing when carbon monoxide and hydrochloric acid may be a byproduct. Avoid skin contact.

Metal colouring oxides Vary in toxicity, some relatively safe (iron), others (cadmium, selenium, chromium, cobalt, vanadium) more toxic. Danger of long-term cumulative effects. Can be absorbed through the skin.

Methylated spirits Denatured alcohol. Used for cleaning off straw hat varnish and other uses. One of the safer solvents, but with some toxic additives.

Nitric acid etch A dilute mixture of nitric acid and water. Acid is always poured into water. Never pour water into acid. Corrosive. When used to etch plates nitrous oxide given off. Ensure adequate exhaust system, wear protective clothing – gloves and goggles.

Oil-based printing media Variety of aromatic solvents. Check health and safety data sheets for specifics. Ensure good ventilation. Avoid skin contact.

Overglaze colours Contain small quantities of lead (in fritted form) and colouring metal oxides.

Plaster of Paris (calcium sulphate) Dust is irritating to the eyes and slightly irritating to the respiratory system. Ensure good ventilation when mixing plaster, wear mask.

Potassium dichromate (potassium bichromate) Used in light sensitive emulsions. Skin contact can cause allergies, irritation and ulceration. Chronic exposure can cause respiratory problems. Suspected carcinogen. Wear protective clothing, gloves and goggles.

Sodium thiosulphate A slow-acting fixing agent used with photographic emulsions. Use with good ventilation. Do not heat (if heated, gives off sulphur oxides, irritating to skin, eyes and respiratory organs).

Straw hat varnish Liquid acid-resistant varnish used in etching, often when making aquatints. Contains solvent. Ensure good ventilation when using.

Turpentine, white spirit, other cleaning solvents A variety of solvents may be used in cleaning and thinning, of varying levels of toxicity. Can be absorbed through the skin, never clean hands with white spirit or turps, use a recognised hand cleaner. Ensure good ventilation during use. 'Odourless' white spirit has toxic aromatic hydrocarbons removed.

Underglaze colours Contain metal oxides in fritted form.

In addition to the above, **ultraviolet light** (UV) emitted from carbon arc lamps, mercury vapour lamps (used for exposing light sensitive emulsions especially screen emulsions) is carcinogenic, and protective measures (goggles, clothing etc.) should be observed when using these.

Car antifreeze has been suggested as a water-based screen printing medium. Antifreeze contains glycol ethers (cellosolves) chemicals which are known to cause damage to the reproductive system. If using these, stop using immediately, and find alternative media.

Computer screens may cause the possible health hazards of headaches, eye strain, stress etc. Ensure comfortable working position, and regular breaks from working in front of screen. Some people have cast doubt about pregnant women using VDUs because of possible increased risk of miscarriages.

Check with your doctor or local Health and Safety Officer.

Cutting rubber or foam stamps with a hot wire or soldering iron is likely to result in toxic fumes. Always ensure good exhaust ventilation.

The very nature of artistic activity means that experimentation is common practice. In all cases the onus is on the artist/ceramist/printmaker to seek specialist advice. Initial information can be obtained from product manufacturers, and further information from Health and Safety officers or other relevant bodies.

Workshop practice

Store all materials safely. Always ensure that products are clearly labelled. Never use old drinks or food bottles or containers unless removal of 'food' labels is possible, and only then, when the result is that the container is unidentifiable as having been used for food or drink storage. Always ensure that hazardous chemicals are stored out of the reach of children.

Install an exhaust/ventilation system if possible. This can serve a number of purposes from glaze spraying, using solvents and etching to removing kiln fumes. Do not attempt to do all these things together, and ensure complete cleaning up before changing operations! Ceramic powders often supplied in plastic-lined paper bags should be transferred to lidded non-breakable plastic containers. Ceramic powders resulting from ceramic activity should be minimised, and mixed with water as soon as possible (offcuts from clay slabs, fettling).

Solvents or products containing solvents should not be kept in glass containers. When not in use, they should be stored in fireproof metal containers. Do not store near kilns, firing equipment or sources of heat. Store below head height to avoid damage to the eyes in the case of a loosely lidded container.

Acids should not be stored in small workshops in concentrated form if at all possible. Store acids below head height. Always wear protective clothing (especially masks, gloves) when handling.

The environment

You are responsible if you pollute the environment as a result of your work and so it is important to minimise the environmental impact of your activities.

In practical terms, when using ceramic materials, you should install a sink trap. This is a tank underneath your sink into which your waste water drains, allowing ceramic materials in suspension to sink before the water is discharged into the drains and sewage system. In practice this means that most glaze materials, metal oxides and underglaze colours are trapped in the tank, forming a sediment on the base. Every so often the tank should be emptied and the sludge or slurry of waste materials disposed of on a recognised landfill site. Some ceramists have suggested drying and firing the sludge so that the ceramic constituents are vitrified. Do not use a sink with a trap when using photographic developers, print media, acids or any kind of solvents.

If using solvents or oil-based products in your studio, never wash these down the sink. In the case of a sink trap, the fumes could build up in an enclosed space and create an explosive or fire hazard. Solvents are polluters of waterways: organic solvents destroy bacteria in sewage plants and septic tanks and prevent proper break down of other wastes. They should be disposed of in an approved manner. Check with the manufacturer, or your local authority, about disposal. There are now recycling schemes for some solvents. Change to water-based inks and processes.

Dispose of used (diluted) acids by further dilution. Do not pour acids into a sink with a sink trap. If in doubt consult relevant local authorities.

Use environmentally friendly cleaning products. Conventional cleaning products can contain petrochemical derivative additives that can be polluting to waterways.

Metal release

Some poisonous metals can be released from fired glazes by contact with acids such as vinegar and fruit juices. There is some evidence that some may be released faster by alkaline foods such as beans or other green vegetables. These include: compounds of lead, antimony, arsenic, barium carbonate, cadmium and selenium oxides, copper, chromium oxide and zinc oxide.

Ceramists making domestic ware using lead glazes or cadmium colours are obliged to have their ware tested for lead and cadmium release in the UK and North America. Ceramists producing ware that might conceivably be used for food use, but who are not sure that the ware is food safe, should label it permanently 'Not for Food Use'. In the USA there are regulations insisting that ware must be rendered unusable or be permanently marked (with decals) 'Not for food use – may poison food'. Alternatively, use lead- and cadmium- free glazes and colours.

Maintaining a sense of proportion

Ensure that you do find out about all the materials you use. Follow Health and Safety advice on use, storage and disposal. Also, always maintain a sense of proportion. Materials are hazardous to the person or environment if not used properly, or if you fail to protect yourself or others from their effects. Few of the ceramic materials normally encountered in this book are hazardous to a degree that demands a 'toxic' labelling. Other chemicals relating to the print or photographic sections of processes may be, and you must make up your own mind about the desirability of using them, in your situation.

Ceramic and print suppliers/information

If ceramic print products do not appear in general ceramic supplier catalogues, contact the supplier and ask if they can get the product you require. In most cases, the companies listed below will be able to help, and, if not, will suggest alternative sources.

American Art Clay Co. Inc., (AMACO)
4717 W 16 Street,
Indianapolis, IN 46222, USA
Tel: +1-(0) 800-374-1600
email: - catalog @ amaco.com
www.amaco.com

European Office:
PO. Box 467
Longton,
Stoke-on-Trent
ST3 7DN, UK
+ 44 (0)1782 399219
Email: - andrewcarter@amaco.uk.co
Ceramic colour supplier, Velvet underglaze screening inks.

K.H. Bailey and Sons Ltd,
Marsh Street,
Stoke on Trent,
Staffs,
ST1 5HH, UK
Tel: +44 (0) 1782 213811,
Fax: +44 (0) 1782 260299
Manufacturer of rubber stamps (open stock designs available), and ceramic transfers, specialists in open stock.

Brittains (TR) Ltd,
Ivy House Paper Mills,
Commercial Road,
Hanley,
Stoke on Trent,
Staffs, ST1 3QS, UK
Tel: +44 (0) 1782 202567,
Fax: +44 (0) 1782 202157

Brittains Tullis Russell Inc.,
Office No 402,
500 Summer Street,
Stamford,
Connecticut 06901, USA
Manufacturers of range of decal papers.

Decorprint UK Ltd
Unit 1, Crabtree Close,
Fenton Industrial Estate,
Fenton, Stoke-on-Trent
Staffs, ST4 2SW, UK
Ceramic transfers.

Elliott & Fay Good,
P.O. Box 105,
Walkerville,
South Australia 5081,
Australia

Folex Limited,
18/19 Monkspath Business Park,
Shirley,
Solihull,
West Midlands
B90 4NY, UK
Tel: +44 (0)121 733 3833
Fax: +44 (0)121 733 3222
e-mail: sales@folex.co.uk
www.folex.co.uk

Folex Imaging
6 Daniel Road
Fairfield,
NJ 07004, USA.
Tel: + 1(0) 973-575-4500 / 800-631-1150
Fax: + 1(0) 973-575-4646
www.folex-usa.com/

Specialist film for use with laser printers and photocopiers.

Heraeus,
Ceramic Colour Division,
Cinderhill Industrial Estate,
Weston Coyney Road,
Longton,
Stoke on Trent,
Staffs, ST3 5LB, UK
Tel: +44 (0)1782 599423
Wide range of lustres sold in quantities from 25g.

G.T. Paper and Packaging
Hedley Terrace,
Lingard Street,
Burslam,
Stoke-on-Trent ST6 2AW, UK
Tel: +44 (0)1782 577328
Suppliers of Pottery Tissue.

High Firing Colour, (Scarva Pottery Supplies),
10 Drummiller Lane,
Scarva,
Co. Armagh,
Northern Ireland BT 63 6BR, UK
Tel: +44 (0) 1762 831864,
Fax: +44 (0) 1762 832135
www.scarvapottery.com/
Ceramic colour suppliers.

Minnesota Clay Company,
8001 Grand Avenue South,
Bloomington,
MN 55420
800 Clay, USA.
www.mm.com/mnclayus/
Stamping pads and other ceramic graphic supplies.

T.N. Lawrence,
208 Portland Road,
Hove, East Sussex BN3 5QT , UK.
Etching, relief printing and screen printing supplies

Lazertran Limited
8 Alban Square,
Aberaeron,
Ceredigion,
SA46 OAD, UK.
Tel: +44 (0) 01545 571149
Fax: +44 (0) 01545 571187
email: mic@lazertran.com
www.lazertran.com
Lazertran transfer paper.

Pamela Morton Ceramics,
22b Holt Road,
Cromer, Norfolk, NR27 9JW, UK
Tel: +44 (0) 1263 512629, fax same.
Producer of custom-made ceramic decals, will do relatively small runs

Potclays,
Brickkiln Lane,
Etruria, Stoke on Trent,
Staffs. ST4 7BP, UK
Tel: +44 (0) 1782 219816,
Fax +44 (0) 1782 286506
www.potclays.com/
General ceramics supplier, clay, glazes, ceramic colours.

Potters Connection,
Anchor Road,
Longton,
Stoke on Trent,
ST3 lJW, UK
Tel: +44 (0)178 259 8729
Fax: + 44 (0)178 259 3054
Ceramic colour supplier.

Potterycrafts,
Campbell Road,
Stoke on Trent, Staffs, ST4 4ET, UK.
Tel: +44 (0) 1782 745000
Fax: +44 (0) 1782 746000
www.potterycrafts.co.uk/
General ceramics supplier, clay, glazes, ceramic colours, some screen printing products.

John Purcell Paper,
15 Rumsey Road,
London,
SW9 0TR, UK
Tel: +44 (0) 1207 737 5199
Supplier of UWET pre-coated decal paper and

media for making inks for water based decal production.

A.J. Purdy,
30, Stort Mill,
River Way,
Harlow, Essex,
CM20 2SN, UK
Tel: +44 (0) 1279 414556
Screen printing supplier.

Sericol Ltd,
Westwood Road,
Broadstairs, Kent,
CT10 2PA, UK
Tel: + 44 (0) 843 67071,
Fax: + 44 (0) 843 604933
(Plus regional offices)
Screen printing supplies.

Photographic suppliers

Silverprint
12b Valentine Place,
London,
SE1 8QH, UK
Tel: +44 (0)207 620 0844,
Fax: + 44 (0)207 620 0129.
Stock a full range of photographic products including materials needed for gum bichromate printing, and Liquid Light, and Silverprint Emulsion.

Rockland Colloid Corporation,
P.O. Box 376,
Piermont, NY 10968, USA
Tel: +1 (0)914 359 5559,
Fax: +1 (0)914 365 6663.
Manufacturer Liquid Light Emulsion

Toyobo Co., Ltd. Graphic Arts Department
2-8, Dojima Hama 2 Chome, Kita-ku, Osaka
530-8230 JAPAN
Tel: + 81-6-6348-3058
Fax: + 81-6-6348-3099
http://www.toyobo.co.jp/e/seihin/xk/print/
Makers and suppliers of Toyobo Photopolymer plates.

UK supplies from:
Nicoll Graphics
at Openshaw International
Woodhouse Road
Todmordan
Lancs, OL14 5TP, UK
+44 (0) 1706 811413

Courses in water-based and ceramic printing processes...

Centre for Fine Print Research
Faculty of Art Media and Design
Bower Ashton Campus
Clanage Road
Bower Ashton
Bristol , BS3 2JT, UK
E.mail. print.research@uwe.ac.uk
Fax +44 (0) 117 344 4824

Magazines and journals

AN for artists,
1st Floor ,
Turner Building,
7-15 Pink Lane
Newcastle Upon Tyne,
NE1 5DW
UK
+ 44 (0)191 241-8000
www.anweb.co.uk

Ceramics Art and Perception
35 William Street,
Paddington,
NSW 2021
Australia.
Tel +61 (0) 29361 5286.
Fax +61 (0) 29361 5402.
www.ceramicart.com.au

Ceramics Monthly
735 Ceramic Place,
PO Boxx 6102
Westerville,
Ohio, 43086-6102.
USA
Tel: +1 (0) 614 523 1660
Fax +1 (0) 614) 891 8960
www.ceramicsmonthly.org

Ceramic Review
21 Carnaby Street,
London,
WIV IPH.
UK
Tel: +44 (0) 207 439 3377.
Fax +44 (0) 207 287 9954.

Contemporary Ceramics in Society
2 Bartholomew Street West,
Exeter,
Devon,
EX2 4TA
UK.
Phone/Fax + 44(0) 139 243-0082
www.ceramic-society.co.uk

Keramik Magazine,
Bensheimer Straße 4a,
D-64653 Lorsch
Germany
Tel: + 49 (0) 62 51 / 58 93 13
Fax: + 49 (0) 62 51 / 58 93 14

Printmaking Today
50 Ferry Street,
Isle of Dogs,
London,
E14 3DT
UK
Tel: +44 (0)207 515 7322,
Journal on prints from fifteenth century to the present day.

Web Sites:

www.rosenthalgroup.net/artweb/mitch
Mitch Lyons monoprinting on paper from clay

www printandclay.net
The Virtual Museum of Print and Clay

www.artcumbria.org/hot_off_the_press.htm
Hot off the Press, ceramics and print web site.

Bibliography

Ceramics

Copeland, Robert, *Blue and White Transfer Printed Pottery*, Shire Publications.

Copeland, Robert, *Spode's Willow Pattern, and other designs after the Chinese*, Studio Vista, 1990.

Cruickshank, Graeme, *Scottish Spongeware*, Scottish Pottery Studies, 1982.

Eames, Elizabeth, *English Medieval Tiles*, British Museum Publications, 1985.

Patricia Failing, *Howard Kottler Face to Face*, University of Washington Press 1995.

Gault, Rosette, *Paper Clay*, A&C Black/UPP, 1998.

Hamer, Frank and Janet, *The Potter's Dictionary of Materials and Techniques*, A & C Black/University of Pennsylvania Press, 1997.

Kelly, Henry E., *Scottish Sponge Printed Pottery*, Lomondside Press, 1993.

Kosloff, Albert, *Ceramic Screen Printing*, Signs of the Times Publishing, 1984.

McConnell, Kevin, *Spongeware and Splatterware*, Schiffer Pub. Co., Pennsylvania, 1990.

Scott, Paul, *Painted Clay, graphic arts and the ceramic surface*, A&C Black/Watson Guptill, 2000

Ed Scott, Paul/Bennett Terry, *Hot off the Press, ceramics and print*, Bellew Publishing/Tullie House 1997.

Print

Faine, Brad, *The Complete Guide to Screen Printing*, Quarto.

Hoskins, Steve, *Water-based Screenprinting*, A&C Black, 2001.

Kinsey, Anthony, *Simple Screen Printing*, Dryad Press.

Mara, Tim, *The Thames and Hudson Manual of Screen Printing*.

Peterdi, Gabor, *Printmaking, Methods Old and New*, Macmillan Publishing Co., Inc. (New York), Collier MacMillan Publishers (London), 1980.

Stobart, Jane, *Printmaking for Beginners*, A&C Black/ Watson Guptill, 2001.

Thompson, George L., *Rubber Stamps and how to make them*, Cannongate Publishing Ltd., 1982.

Turner, Silvie, *A Printmaker's Handbook*, Estamp, 1989.

Westley, Ann, *Relief Printmaking*, A&C Black/ Watson Guptill, 2001.

Whale, George & Naren Barfield, *Digital Printmaking*, A&C Black, 2001.

Health and safety

Challis, Tim, *Printsafe, A Guide to Safe, Healthy and Green Printmaking*, Estamp, 1990.

Challis, Tim, and Gary Roberts, *Artists Handbooks 2 Health and Safety, Making Art, Avoiding the Dangers*, AN Publications, 1990.

McCann, Michael, *Health Hazards Manual for Artists*, Nick Lyon Books, USA, 1985.

Rossol, Monona, *The Artist's Complete Health and Safety Guide*, Allworth, Press, New York, 1990.

Photographic

Frederick, Peter, *Creative Sunprinting*, Focal Press.

Nadeau, Luis, *Gum Dichromate*.

Nadeau, Luis, *Modern Carbon Printing*.

Silverprint Catalogue. Specialist photographic suppliers, catalogue contains details of chemicals and processes for Gum Bichromate printing and Liquid Light/Silverprint Emulsion. From Silverprint Ltd., 12 Valentine Place London, SE1 8QH.

Wade, Kent E., *Alternative Photographic Processes*, Morgan and Morgan.

Index